BIOMOLECULES

BIOMOLECULES

T. DEVASENA

Visiting Faculty
Centre for Nanoscience and Technology
Anna University, Chennai
Tamil Nadu

MJP PUBLISHERS

Cataloguing-in-Publication Data

Devasena, T (1975 –)
    Biomolecules / by T. Devasena. –
Chennai : MJP Publishers, 2010
    xvi, 264p. ; 21 cm.
    Includes glossary, references and index.
    ISBN 978-81-8094-079-8 (pbk.)
    1. Molecules 2. Biology, Molecules
I. Title
    572.82 dc22 DEV MJP 067

ISBN 978-81-8094-079-8
© Publishers, 2010
All rights reserved
Printed and bound in India

MJP PUBLISHERS
47, Nallathambi Street
Triplicane
Chennai 600 005

Publisher : J.C. Pillai
Managing Editor : C. Sajeesh Kumar
Marketing Manager : S.Y. Sekar
Project Editor : P. Parvath Radha
Acquisitions Editor : C. Janarthanan
Editorial Team : B. Ramalakshmi, N. Yamuna Devi,
M. Gnanasoundari, Lissy John, R. Magesh
CIP Data : Prof. K. Hariharan, Librarian
RKM Vivekananda College, Chennai.

PREFACE

Biologists must constantly keep in mind that what they see was not designed, but rather evolved.

Francis Crick
What Mad Pursuit (1990), 138.

This book is suited for an introductory course in biomolecules for students of biochemistry, biotechnology, nanotechnology, microbiology, pharmacy, zoology and other life sciences at undergraduate and postgraduate levels. The primary goal is to help students understand cell biology and the various molecules in the biological system like nucleic acids, proteins, enzymes, carbohydrates, lipids and water. Vitamins which are indispensable for the growth and function of the biological system is also included as a chapter.

The introductory chapter gives general concepts of various biomolecules. This is followed by nine chapters each one discussing a specific biomolecule in detail.

The cell is the fundamental unit of the biological system in which all the biomolecules are synthesized, actuated and metabolized. Chapter 1 focuses on the structure and function of animal, plant, yeast, bacterial and viral cells. The cell division process and reproductive phases are also discussed with suitable diagrams and flowcharts.

The central dogma of life starts with DNA or gene and continues with RNA. The structure, function and metabolism of DNA and RNA along with replication of DNA and transcription of RNA are all detailed in chapter 2. Amino acids and proteins are discussed in chapter 3. This chapter covers the classification of amino acids and proteins, tests for identification of amino acids and proteins, structural diversity of proteins and metabolism of amino acids. A separate topic on pharmaceutical proteins gives information about the synthesis of pharmaceutical proteins via genetic engineering and various applications of these proteins. Enzymes — the most exciting area in the field of biomolecules — are highlighted in chapter 4. The structure of enzymes, models for enzyme — substrate interaction, mechanism of enzyme action and applications of enzymes are featured in this chapter.

Chapter 5 deals with carbohydrates. The classification of carbohydrates, and the structure and metabolism of carbohydrates are discussed. Blood group antigen is one of the important topics covered in this chapter. In chapter 6, fatty acids and lipids are discussed. This chapter reveals the types of lipids, properties of lipids and metabolism of lipids. The process that converts the metabolic end product of biomolecules into energy is covered in Chapter 7. This chapter deals with the pyruvate dehydrogenase reaction, TCA cycle, electron transport chain and oxidative phosphorylation and formation of ATP.

Water, the vital nutrient of life and the most important of all biomolecules, is discussed in chapter 8. This chapter explains the structure and functions of water. Vitamins are not synthesized in the body but still they are essential for the growth of the body. Vitamins are also precursors for coenzymes which catalyse biochemical reactions and help proper functioning of biological systems. Thus, in chapter 9, the sources, structure, function and deficiency diseases of fat-soluble and water-soluble vitamins are discussed.

Each chapter introduces a particular relevant scientist giving his/her achievement in order to inspire and motivate the students and kindle their interest in going deep into the subject. Each chapter ends with review questions and MCQs for self assessment.

T. Devasena

ACKNOWLEDGEMENTS

I am indebted to Dr. Mannar Jawahar, Vice Chancellor, Anna University Chennai, for encouraging and supporting me to write this book.

I acknowledge Dr. R. Jayavel, Director, Centre for Nanoscience and Technology, Anna University Chennai, Chennai for the help, suggestion and encouragement he rendered throughout the time of preparation of this book. I am grateful to Dr. Aruna Sivakami, Vice Chancellor, Mother Teresa Women's University, for her encouragement and support.

I am thankful to Dr. S. Sundaram, Assistant Professor, Department of Indian History, University of Madras, Chennai, for his constant encouragement and moral support to write this book.

I thank Mr. N. Krishna Chandar, National Doctorate Fellow, Centre for Nanoscience and Technology, Anna University Chennai, for his technical assistance in preparing the manuscript.

I also thank N. Sushma, Head, Department of Biochemistry, Vel's University, Chennai, for her kind help during the preparation of the manuscript.

I finally thank the editorial team of MJP Publishers for their active cooperation and effective coordination and for the speedy publication of this book.

T. Devasena

CONTENTS

INTRODUCTION   **1**

Types of Biomolecules   1

Proteins   1

Enzymes   2

Nucleic Acids   3

Carbohydrates   3

Lipids   4

Vitamins   5

Water   6

1. CELL BIOLOGY   **7**

EUKARYOTES   **8**

Animal Cell   **10**

Shape and Size   10

Subcellular Organelles   10

     Cytoplasm   11

     Nucleus   12

     Endoplasmic Reticulum   13

     Golgi Complex/Golgi Stack   14

     Lysosomes   15

     Peroxisomes   15

     Ribosomes   16

     Mitochondria   16

     Centrosome   17

     Vacuole   17

     Membrane (Cell Membrane/Plasma Membrane)   18

     Cytoskeleton   22

Cell Division	**25**
G1 Phase—The First Gap Phase	27
S Phase—The Synthetic Phase	27
G2 Phase—The Second Gap Phase	27
M Phase—The Mitotic Phase	27
Mitosis	27
Cytokinesis	29
Meiosis	29
Control of Cell Cycle (Regulation of Cell Cycle)	31
Checkpoints	31
Components in Cell Cycle Regulation	32
Fungi	**36**
Structure	36
Fungal Dimorphism	37
Y ↔ M Shift	37
M ↔ Y Shift	38
Reproduction in Fungi	38
Asexual Reproduction	39
Sexual Reproduction	41
Algae	**42**
Structure	42
Encystment and Excystment	43
Reproduction in Algae	43
PROKARYOTES	**45**
Bacterial Cell	**45**
Shape and Size	45
Structure	45
Structures Forming the Surface and Internal Part of the Cell	46
Structures Forming the Outer Part of the Cell	49
Gram-Positive and Gram-Negative Bacteria	52
Reproduction in Bacteria	54

Asexual Reproduction 54

Reproduction by Exchange
of DNA/Sexual Reproduction 57

VIRUS **57**

Structure of Virus 57

Reproduction in Virus 59

Review Questions 60

2. **NUCLEIC ACID** **63**

Nucleotides 63

Pentose Sugar 63

Nitrogenous Base 64

Phosphate Group 65

STRUCTURE OF NUCLEIC ACIDS **65**

Structure of DNA 65

Primary Structure of DNA 65

Secondary Structure of DNA 67

Tertiary Structure of DNA 68

Structure of RNA 68

Primary Structure of RNA 68

Secondary Structure of RNA 68

Tertiary Structure of RNA 68

Replication of DNA 70

Requirements 71

Mechanism of Replication 72

Transcription 75

Requirements 76

Mechanism of Transcription 76

Post-transcriptional Modification of RNA 78

Translation 83

Codons 84

Components Needed for Translation 84

Mechanism of Translation 84

Elongation 85

Termination	89
Post-translational Modification	90
Metabolism of Nucleotides	93
De novo Pathway for the Synthesis of Purine Nucleotides	94
Salvage Pathway for the Synthesis of Purine Nucleotides	96
De novo Pathway for the Synthesis of Pyrimidine Nucleotides	96
Salvage Pathway for the Synthesis of Pyrimidine Nucleotides	98
Degradation of Nucleotides	98
Review Questions	*99*
3. PROTEINS	**101**
Amino Acids — Building Blocks of Proteins	101
Charge Distribution in Amino Acids	105
Essential and Nonessential Amino Acids	106
Non-standard Amino Acids	106
Linking of Amino Acids	107
Properties of Amino Acids	108
Physical Properties	108
Chemical Properties	108
Colour Reactions for Specific Amino Acids	110
Structure of Proteins	111
Primary Structure	113
Secondary Structure	113
Tertiary Structure	114
Quaternary Structure	115
Interactions in Protein Structure	115
Covalent Interaction	115
Noncovalent Interaction	117
Types of Proteins	119
Types of Proteins based on Chemical Composition	119

Types of Proteins
  based on Biological Function 119
Protein Structure Determination 120
  Determination of Primary Structure
    (Determination of Amino Acid Composition) 120
  Determination of Secondary
    and Tertiary Structure 124
Pharmaceutical Proteins 125
  Steps in Producing Pharmaceutical
    Proteins by Genetic Engineering 125
  Examples for Pharmaceutical Proteins 129
Metabolism of Amino Acids 130
  Catabolism or Degradation
    of Amino Acids 130
  Biosynthesis of Amino Acids 132
  Review Questions 134

4. **ENZYMES** **137**
Ribozymes and Isozymes 137
Enzyme Activity 138
Properties of Enzymes 138
Factors Influencing Enzyme Activity 140
  Effect of Substrate Concentration 140
  Effect of Temperature 141
  Effect of pH 141
Structure of Enzymes 142
Models for Enzyme–
  Substrate Reaction 143
  Fischer's Lock-and-Key Model 143
  Koshland's Induced Fit Model 144
Mechanism of Enzyme Action 144
Rate Equation for Enzyme-catalysed
  Reaction — Michaelis–Menten Equation 145
Enzyme Inhibition 146
Coenzymes and Cofactors 148

Applications of Enzymes 149

Review Questions 150

5. **CARBOHYDRATES** **153**

Functions of Carbohydrates 153

Nomenclature of Sugars based on Number
of Carbon Atoms 154

Classification of Carbohydrates 155

Classification of Carbohydrates
based on Functional Groups 155

Classification of Carbohydrates
based on Number of Sugar Residues 155

Structure of Carbohydrates 156

Open Chain Form 156

Cyclic Form 161

Ring Form 162

Glycoside Linkage 163

Polysaccharides 164

Starch 164

Glycogen 166

Cellulose 166

Chitin 166

Inulin 167

Carbohydrate Derivatives 167

Sugar Esters 167

Sugar Acids 168

Sugar Alcohol 168

Amino Sugars 169

Deoxy Sugars 169

Heteropolysaccharides 170

Peptidoglycan 170

Glycosaminoglycan (GAG) 171

Proteoglycan 171

Glycoproteins 171

Functions of Glycoproteins   172
Examples of Glycoproteins   173
Metabolism of Carbohydrates   175
   Glycolysis   175
   Gluconeogenesis   177
   Glycogenesis   180
   Glycogenolysis   182
   Pentose Phosphate Pathway   183
Review Questions   *184*

6. LIPIDS   **187**
Essential Fatty Acids   188
Nomenclature of Fatty Acids   188
Functions of Lipids   189
Physical Properties of Lipids   189
Chemical Properties   190
Classification of Lipids   192
   Simple Lipids   192
   Compound Lipids   195
   Derived Lipids   199
Metabolism of Fatty acids   202
   Biosynthesis of Fatty Acids   202
   Synthesis of Unsaturated Fatty Acid   203
   Fatty Acid Oxidation/ Fatty Acid Degradation   203
Review Questions   *206*

**7. ELECTRON TRANSPORT CHAIN
AND OXIDATIVE PHOSPHORYLATION**   **209**
Pyruvate Dehydrogenase Reaction   210
TCA Cycle   210
Electron Transport Chain   213
Oxidative Phosphorylation   215
Uncouplers   216
Review Questions   *217*

8. WATER   **219**
   Structure of Water   219
   Hydrogen Bonding in Water   220
   Properties of Water   221
   Functions of Water   222
      Stabilization of Proteins by Water   222
      Stabilization of Nucleic Acids by Water   223
      Binding of Biomolecules   225
   Review Questions   226

9. VITAMINS   **229**
   Classification of Vitamins   230
   FAT-SOLUBLE VITAMINS   **230**
   Vitamin A (Antixerophthalmic vitamin)   230
   Vitamin D (Antirachitic vitamin)   235
   Vitamin E (Antisterilitic vitamin)   236
   Vitamin K (Coagulation vitamin or
      antihaemorrhagic vitamin)   238
   WATER-SOLUBLE VITAMINS   **239**
   Vitamin B_1 (Thiamine)   239
   Vitamin B_2 (Riboflavin)   240
   Vitamin B_3 (Niacin)   241
   Vitamin B_5 (Pantothenic acid)   243
   Vitamin B_6 (Pyridoxine)   244
   Vitamin B_7 (Biotin)   245
   Vitamin B_9 (Folate)   246
   Vitamin B_{12} (Cyanocobalamin)   247
   Vitamin C (Ascorbic acid)   249
   Review Questions   250
Glossary   253
References   259
Index   261

HERMANN STAUDINGER
Awarded Nobel Prize in Chemistry for his discoveries in the field of macromolecular chemistry.

Biomolecules are organic molecules produced by living organisms. They are primarily composed of carbon, hydrogen, nitrogen and oxygen.

TYPES OF BIOMOLECULES

Biomolecules are broadly divided into three types.

Monomers — amino acids, nucleotides, monosaccharides.

Polymers — proteins, enzymes, carbohydrates, nucleic acids, lipids.

Small molecules — water, vitamins (should be supplied in the diet but needed for proper body function).

PROTEINS

Proteins constitute 10–30% of the cell mass. The primary elements constituting proteins are carbon, nitrogen, hydrogen and oxygen and sometimes sulphur. Amino acid monomers are linked by peptide bonds to form peptide chain which in turn forms functional proteins. Functional proteins include enzymes, haemoglobin, hormones and antibodies. The primary structure of a protein is the amino acid sequence of the peptide chain. The secondary structure is formed by hydrogen bonding of the peptide chain. Folding of proteins into layers, bundles or globules form the three-dimensional tertiary structure. Spatial arrangements of different subunits of proteins (tertiary structure) result in quaternary structure.

ENZYMES

Enzymes are specialized protein molecules facilitating most of the body's metabolic processes such as, catalysing biochemical reactions, supplying energy, digesting foods, purifying blood and ridding the body of waste products, etc. They are divided into two main groups: metabolic enzymes and digestive enzymes. They assist in fighting ageing, weight loss, lowering cholesterol, cleaning the colon, breaking down fats, strengthening the immune system, improving mental capacity, detoxifying the body, building muscles from protein, eliminating carbon dioxide from the lungs, etc. No matter how healthy we eat or how many vitamins and minerals we load into our diets, they **would not be worth anything without enzymes**. The most well known and important enzymes are the **digestive enzymes** which are divided into seven categories: lipase (breaking down fat), protease (breaking down proteins), cellulase (breaking down fibre), amylase (breaking down starch), lactase (breaking down dairy), sucrase (breaking down sugars) and maltase (breaking down grain). Enzymes can be obtained from fruits and vegetables like mangoes, avocados, papaya, asparagus, etc.

NUCLEIC ACIDS

Nucleotides are linked by phosphodiester bond to form nucleic acid polymer. It is basically made up of base, sugar, and phosphate. Depending on the sugar nucleic acid is divided into deoxyribonucleic acid or DNA (which possesses deoxyribose sugar) and ribonucleic acid or RNA (which possesses ribose sugar). DNA exists in double helical form. RNA exists in stem loop form and clover leaf form. Compounds which are similar to nucleic acids are called nucleic acid analogues or artificial nucleic acids. Examples are peptide nucleic acid (PNA), Morpholino and locked nucleic acid (LNA), as well as glycol nucleic acid (GNA) and threose nucleic acid (TNA). Each of these is distinguished from naturally occurring DNA or RNA by changes in the backbone of the molecule. Artificial nucleic acids are used in molecular biology and genetic engineering research.

CARBOHYDRATES

Carbohydrates are organic molecules with the general formula of CHO in 1 : 2 : 1 ratio. Monosaccharides are linked by glycosidic bond to form carbohydrates. Although carbohydrates constitute only 1 to 2 per cent of cell mass, they provide the raw fuel for cellular energy production. Carbohydrates are classified according to molecular size and solubility. In general, the smaller molecules are more soluble than the larger ones. Monosaccharides include glucose, fructose, galactose, deoxyribose, and ribose. Monosaccharides (comprised of single sugar unit) are the simplest carbohydrates. Glucose, a 6-carbon sugar (hexose) is the sugar in our blood. Fructose, the sugar that sweetens fruit, and lactose, the sugar found in milk, have the same chemical formula as glucose and are therefore isomers of glucose. Glucose can have a straight chain of carbon atoms or, more commonly, form a ring structure. Two other 5-carbon sugars or monosaccharides (called pentose) used in nucleic acid synthesis are deoxyribose and ribose.

Two monosaccharides are joined together by dehydration synthesis to form a disaccharide molecule. Disaccharides include sucrose, lactose, and maltose.

Glucose + Fructose = Sucrose (table sugar) + Water

Glucose + Galactose = Lactose (milk sugar) + Water

Glucose + Glucose = Maltose (malt sugar) + Water

Polysaccharides include starch, cellulose, and glycogen. These long, chainlike polymers make ideal storage products due to their insolubility. Starch is the storage molecule synthesized from glucose by plants. Cellulose, which is also synthesized by plants for cell wall construction, is indigestible because we lack enzymes for it. Cellulose provides fibre to promote peristalsis. Glycogen is the storage polysaccharide in animals. It is stored in liver.

LIPIDS

All lipids are hydrophobic molecules. Fats, oils, waxes, phospholipids and steroids are lipids of biological significance. Fats and oils are made from two kinds of molecules: glycerol (a type of alcohol with a hydroxyl group on each of its three carbons) and three fatty acids joined by dehydration synthesis. Phospholipids are made from glycerol, two fatty acids, and (in place of the third fatty acid) a phosphate group with some other molecule attached to its other end. Waxes are esters of fatty acids with long-chain monohydric alcohols (one hydroxyl group). Steroids have structures totally different from the other classes of lipids. The main feature of steroids is the ring system of three cyclohexanes and one cyclopentane in a fused ring system. Steroids include such well-known compounds as cholesterol and sex hormones. Nearly all of the energy needed by the human body is provided by the oxidation of carbohydrates and lipids. While carbohydrates provide a readily available source of energy, lipids function primarily as an energy reserve. The amount of lipids stored as an energy reserve far exceeds the energy stored as glycogen since the human body is simply not capable of storing as much glycogen compared to lipids. Lipids yield 9 kcal of energy per gram while carbohydrates and proteins yield only 4 kcal of energy per gram. Lipids or fats are stored in cells throughout the body principally in special kinds of connective tissue called adipose tissue or depot fat. While, many cells contain phospholipids in the bilayer cell membranes, adipose tissue cells consist of fat globules of triglycerides which may occupy as much as 90% of the cell volume. In addition to energy storage, depot fat provides a number of other functions. Fat serves as a protective cushion and provides structural support to help prevent injury to vital organs such as the heart, liver, kidneys, and spleen. Fat insulates the body from heat loss and extreme temperature changes. At the same time, fat deposits under the skin may be metabolized to generate heat in response to lower skin temperatures. Since lipids are not soluble in blood, they are transported as lipoproteins after reaction with water-soluble proteins in the blood.

VITAMINS

Vitamin is an organic substance that acts as a coenzyme and/or regulator of metabolic processes. There are 13 known vitamins, most of which are present in foods or supplements; some are produced within the body. Vitamins are crucial for many bodily functions. Vitamins are grouped into two categories: Fat-soluble vitamins are stored in the body's fatty tissue. Water-soluble vitamins must be used by the body right away. Any leftover water-soluble vitamins leave the body through the urine. Vitamin B_{12} is the only water-soluble vitamin that can be stored in the liver for many years.

WATER

Water is a vital nutrient. The human body can last weeks without food, but only days without water. The total body mass is made up of approximately 80% water. Water forms the basis of blood, digestive juices, urine and perspiration and is contained in lean muscle, fat and bones. As the body can't store water, we need fresh supplies everyday to make up for losses from lungs, skin, urine and faeces. The amount we need depends on our metabolism, the weather, the food we eat and our activity levels. Apart from being a vital nutrient, water is an excellent solvent. Although not all things dissolve in water, it is often referred to as the "universal solvent". Water is a polar molecule and its geometry is such that the electrons are not uniformly distributed throughout the molecule. So the end of the molecule with greater electron density is slightly negative and the other slightly positive. This is one reason why water boils at such a high temperature — its molecules stick together tightly and it takes a lot of energy to break them apart. Water generally dissolves other substances that are also polar, but not non-polar substances like oil.

All these biomolecules are discussed in detail in the subsequent chapters with reference to structure, function, metabolism and biological significance.

CELL BIOLOGY

TIM HUNT
Discovered key regulators of the cell cycle.

Cell is the basic structural and functional unit of living organisms. Cells assemble to form tissues. Tissues are organized into organs and organs into organ systems which together constitute an organism.

Cells → Tissues → Organs → Organ systems → Organisms

Organisms can basically be classified into prokaryotes (e.g. bacteria), eukaryotes (e.g. animals and plants) and viruses. Prokaryotes differ from eukaryotes in various aspects. Table 1.1 shows the differences between prokaryotes and eukaryotes. Examples for these three types of organisms are discussed in this chapter.

Table 1.1   Differences between prokaryotes and eukaryotes

Characteristic features	Prokaryotes	Eukaryotes
Subcellular organelles	Not present	Present
Genetic information	DNA is naked	DNA is enclosed in a membrane-bound nucleus.
Cell division	Occurs by binary fission	Occurs by mitosis and meiosis.
Protein synthesis	Transcription and translation are coupled in nucleoid itself	Transcription occurs in nucleus; translation occurs in cytoplasm.
Metabolism	Aerobic or anaerobic	Aerobic
Respiratory enzymes	Present in plasma membrane	Embedded in mitochondria.
Cell wall	Present	Absent in animals cells. Present only in plant cells.
Exocytosis and endocytosis	Cannot occur.	Can occur.

EUKARYOTES

The animal cell is the best example for a eukaryotic cell. Animal cells are mostly spherical. The cell is made up of an outer membrane and an inner cytoplasm. Many membrane-bound subcellular organelles are embedded in the cytoplasm. Plant cells are made up of a thick cell wall that covers the membrane. The structures of the plant cell and animal cell are represented in Figure 1.1. The differences between the animal cell and plant cell are shown in Table 1.2.

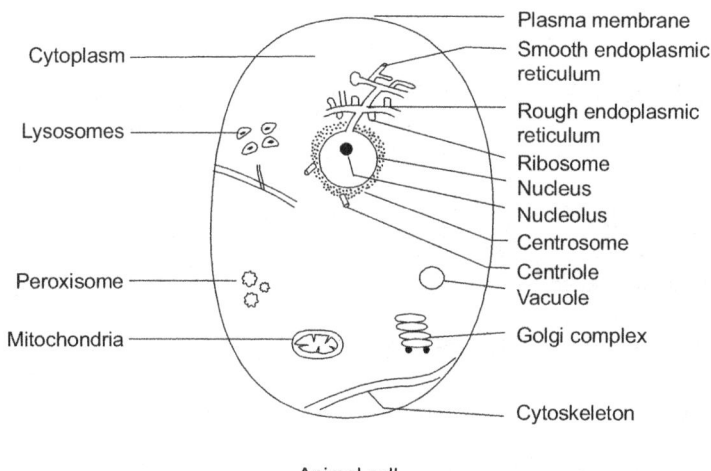

Animal cell

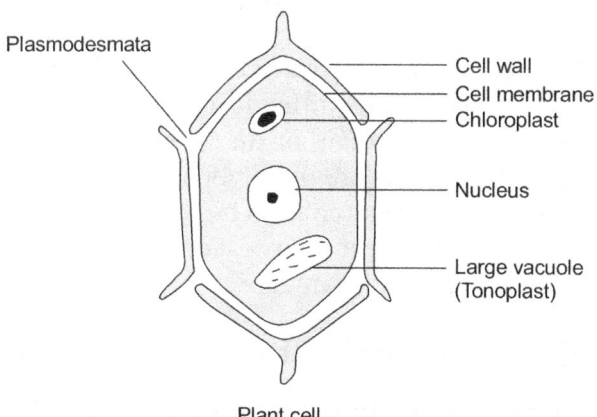

Plant cell

Figure 1.1 Structure of animal cell and plant cell

Table 1.2 Differences between animal cell and plant cell

Features	Animal cell	Plant cell
Cell size	Small	Large
Cell wall	Absent	Present
Vacuole	Small	Large
Chloroplast	Present only in a few cells like *Euglena*	Present in all types of plant cells
Centriole	Present	Absent
Cytokinesis (Cytoplasmic division)	Takes place by constriction.	Takes place by plate formation.

ANIMAL CELL

SHAPE AND SIZE

Diverse shapes exhibited by eukaryotic cells are shown in Figure 1.2. Eukaryotic cells are spherical. Some are cuboidal, polygonal, cylindrical, oval, elongated or flattened. For example, muscle cells are elongated, epithelial cells are flattened or disc-shaped. Sperm cells are elongated, egg cells are oval, hepatocytes are lobular and nerve cells are branched. The size of cells ranges between 1 μm and 175 μm. Largest cell is the ostrich egg cell which is 175 mm in diameter.

SUBCELLULAR ORGANELLES

Small membrane-bound organelles embedded in the cytoplasm are called subcellular organelles. The various subcellular organelles are discussed below.

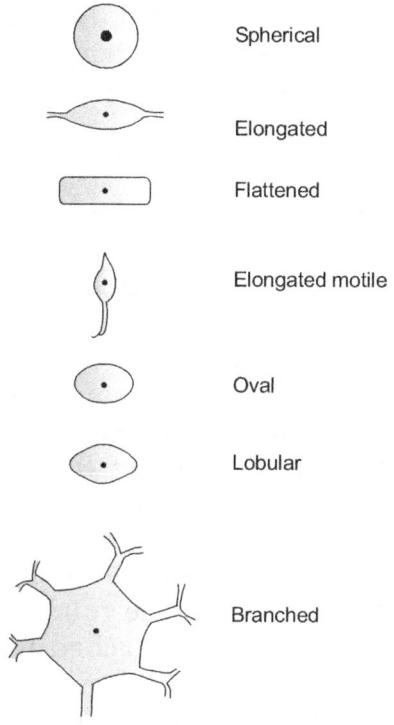

Spherical

Elongated

Flattened

Elongated motile

Oval

Lobular

Branched

Figure 1.2   Shapes of cells

Cytoplasm

Cytoplasm is an amorphous, translucent and homogeneous colloidal liquid existing between the nucleus and the membrane. It is made up of inorganic constituents (water, sodium salts, potassium salts and metals), and organic constituents (carbohydrates, lipids, proteins, nucleoproteins). It has living structures called organelles and non-living structures called **paraplasm** or **deutoplasm** or **inclusions** (e.g. pigments). The peripheral part of the cytoplasm which is nearer to the membrane is non-granular and viscous and is called **ectoplasm** or **plasmagel**. The inner part of the cytoplasm which is nearer to the nucleus is granular and less viscous and this part is called the **endoplasm**. The cytoplasm is the site for many enzymes responsible for metabolic pathways. It is the site for protein synthesis.

Nucleus

The nucleus is a more or less spherical organelle. The structure of the nucleus is shown in Figure 1.3. It has a double membrane system called **nuclear envelope** which is not continuous but interrupted by small holes called **nuclear pores.** The nuclear pore accommodates the nuclear pore complex which contains protein granules. The inner membrane has a leaflike structure called nuclear lamina which is a network of fibres. It has three types of proteins called **lamins** which play a role in mitosis (cell division). The nuclear envelope is the dynamic part of the nucleus. The outer membrane of the envelope is attached to another organelle called the rough endoplasmic reticulum which in turn is attached to the smooth endoplasmic reticulum. A small spherical structure called nucleolus is present in the inner part of the nucleus. The nucleus stores nucleic acids. Two types of nucleic acids are found — DNA and RNA (deoxyribonucleic acid and ribonucleic acid).

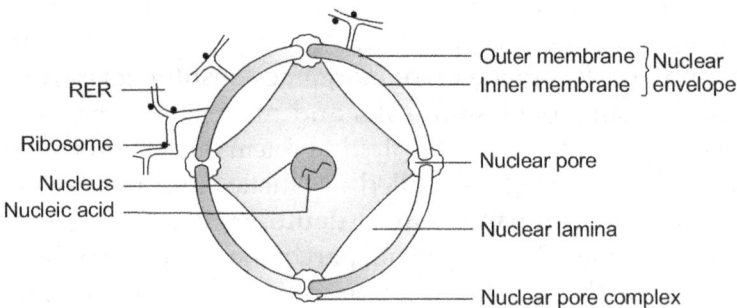

Figure 1.3   Structure of nucleus

Central dogma of life Within the nucleus, DNA molecules are replicated to form daughter DNA by the process of **replication**. Information coded in DNA is converted to mRNA by the process called **transcription**. In the cytoplasm, information encoded in RNA is converted into protein by the process called the **translation**. This sequential process is called central dogma of life and it can be represented as follows:

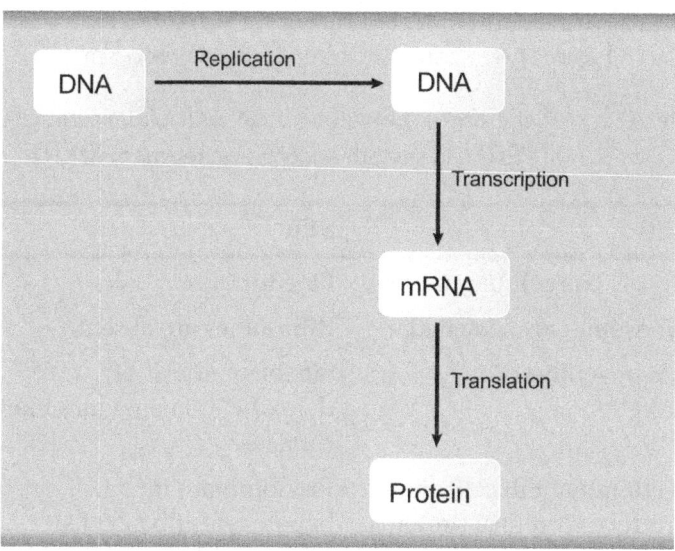

Endoplasmic Reticulum

The endoplasmic reticulum (ER) is a highly convoluted membranous sac. It is made up of interconnected tubules. It forms about 50% of the entire cell membrane. It is continuous with the outer membrane of the nucleus (Figure 1.4). There are two types of ER: rough endoplasmic reticulum (RER) and smooth endoplasmic reticulum (SER). Differences between RER and SER are given in Table 1.3.

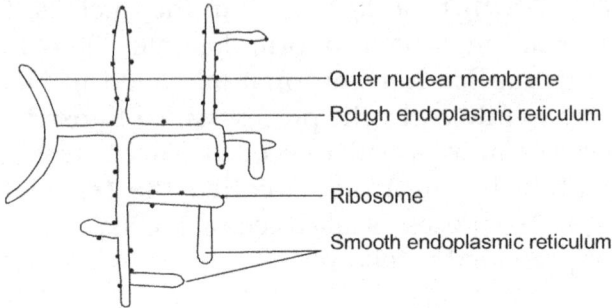

Outer nuclear membrane
Rough endoplasmic reticulum

Ribosome

Smooth endoplasmic reticulum

Figure 1.4   Structure of endoplasmic reticulum

Table 1.3   Differences between rough endoplasmic reticulum (RER) and smooth endoplasmic reticulum (SER)

RER	SER
Highly convoluted tubules.	Fine tubules.
Ribosomes are attached.	Ribosomes are absent.
Site of synthesis of proteins.	Site of synthesis of detoxification enzymes like catalase.
Predominant in cells that produce secretory proteins.	Predominant in hepatocytes (liver cells) where lipid metabolism takes place.

Golgi Complex/Golgi Stack

Golgi complex refers to the membranous stacklike structure formed by flattened discs. It is found associated with small membrane-bound vesicles called Golgi vesicle. In plant cells, it is referred to as **dictyosomes**. It is the **principal director** for the intracellular movement of macromolecules like lipids, proteins, carbohydrates, etc.

It is involved in tagging, i.e., adding small molecules to large molecules for targeting into specific organelles. For example, the Golgi stack is responsible for adding phosphate group to mannose sugar and the phosphorylated mannose is targeted to lysosomes. Similarly proteins can be tagged with oligosaccharides.

Lysosomes

Lysosomes are very small, either spherical or irregular in shape. They are membrane-bound saclike structures, which store an array of hydrolytic enzymes collectively called **acid hydrolase**. These enzymes are involved in digestion of macromolecules. For example, RNase is involved in the digestion of RNA; acid phosphatase is involved in digestion of phosphates; DNase is involved in the digestion of DNA; proteases (cathepsin, collagenase, peptidase) are involved in the digestion of proteins; galactosidase, hexosaminidase, lysozyme and hyaluronidase are involved in the digestion of carbohydrates and esterase. Phospholipase is involved in degrading lipids. Lysosomes are called **suicidal bags** of the cell.

Peroxisomes

Peroxisomes are small structures similar to lysosomes. They store detoxification enzymes and help in detoxification by removing toxic substances. For example, catalase (CAT) detoxifies hydrogen peroxide which is otherwise toxic.

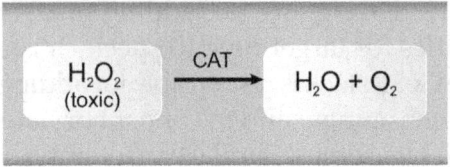

Ribosomes

Ribosomes are organelles responsible for protein synthesis. There are two types of ribosomes. Monosomes which exist as single units associated with ER. They are responsible for synthesis of secretory proteins (proteins secreted out of the cell) and organelle-targeted proteins (proteins transported into specific organelles within the cell). Polyribosomes or polysomes which exist as multiple units are associated with mRNA and are responsible for the synthesis of cytosolic proteins (Figure 1.5).

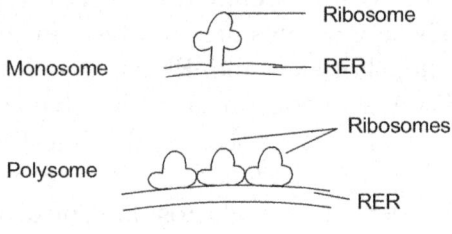

Figure 1.5   Monosome and polysome

Mitochondria

Mitochondria are the organelles responsible for respiration of cells and energy production. They are called the **powerhouse** or **power plant** of the cell. The structure of mitochondrion is shown in Figure 1.6. It consists of an outer membrane which is smooth and an inner membrane which is folded to form structures called **cristae**. The inner membrane is made up of enzymes involved in electron transport chain. The inner part of the mitochondrion is called matrix which hosts DNA, ribosomes and enzymes. Mitochondrial DNA replicates independently and its genetic code is different from that of nuclear DNA.

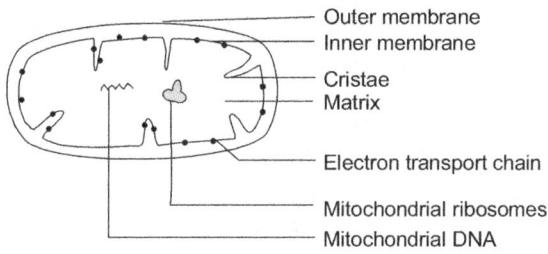

Figure 1.6   Structure of a mitochondrion

Centrosome

The dense region of cytoplasm that surrounds the nucleus is called centrosome. It is involved in cell division (Figure 1.7). It has a pair of small organelles, the centrioles, each made up of a ring of nine groups of microtubules. The two centrioles are arranged such that one is perpendicular to the other.

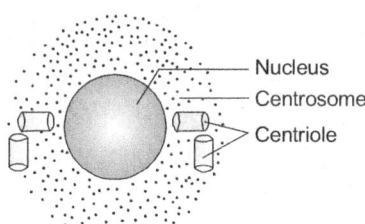

Figure 1.7   Centrosome

Vacuole

Vacuoles are hollow structures filled with liquid. They may be spherical or nonspherical. They are responsible for storage of cell liquids and maintain the internal pressure of the cell. Plant vacuoles are larger than animal vacuoles and are bound by a well-defined membrane called **tonoplast.**

Membrane (Cell Membrane/ Plasma Membrane)

Membrane is the external protective covering of the cell. It acts as a thin barrier that separates the intracellular and extracellular compartments. It is porous, elastic, ultrathin and semipermeable in nature. It acts as a mechanical support to the cell and segregates intracellular events.

Composition of membrane   Membrane is made up of the following components:

- Lipids — phospholipids, glycolipids, cholesterol
- Carbohydrates — glycolipids, glycoproteins
- Proteins — integral proteins and peripheral proteins
- Enzymes — soluble enzymes and insoluble enzymes (membrane-bound enzymes).

Lipids   Membrane lipids are amphipatheic in nature, i.e., they have a water-soluble (hydrophilic) part called **head** and a water-insoluble (hydrophobic) part called **tail**. Lipids are arranged in the form of a bilayer with hydrophilic head facing outward and the hydrophobic tail facing inward (Figure 1.8). The lipid bilayer of the membranes undergoes two types of movements: lateral diffusion and translational movement.

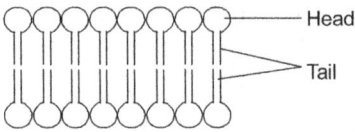

Figure 1.8   Lipid bilayer

Proteins Some of the membrane proteins span through the entire lipid bilayer and they are called integral proteins or intrinsic proteins. Some of the proteins are loosely bound to the lipid bilayer and they do not span the entire lipid bilayer. These proteins are called peripheral proteins or extrinsic proteins (Figure 1.9). Table 1.4 shows the differences between peripheral and integral proteins.

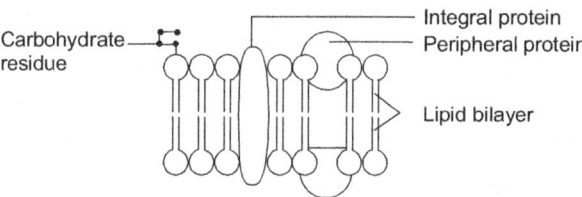

Figure 1.9 Integral proteins and peripheral proteins

Table 1.4 Differences between peripheral and integral proteins

Property	Peripheral proteins	Integral proteins
Dissociation from membranes	Mild treatment is essential, e.g. chelating agents.	Strong treatment is essential, e.g. detergents and organic solvents.
Location	Surface of the membrane	Spans the entire bilayer.
Association with lipids	Free of lipids	Strongly associated with lipids.
Examples	Enzymes involved in electron transport chain. Spectrin present in red blood cell membrane.	Drug and hormone receptors.

Functions of membrane proteins   Membrane proteins provide mechanical support to the membrane and act as a channel or carriers for transport of molecules across the membrane. They also serve as membrane-bound enzymes and receptors.

Carbohydrates   Carbohydrates form a minor part of the membrane. It is found in association with lipids and proteins.

Molecular architecture of membrane   Molecular architecture of membrane can be understood by picturing four proposed models:

1. Lamellar model
2. Micelle model
3. Protein crystal model
4. Fluid mosaic model (widely accepted model).

These models are shown in Figure 1.10.

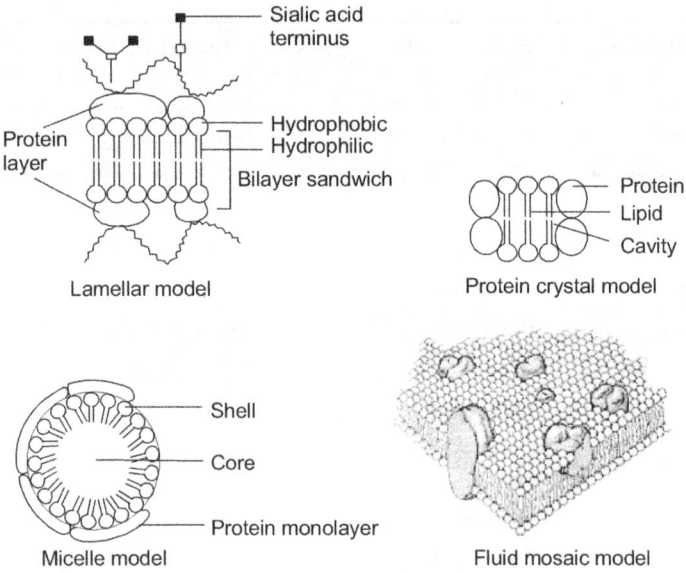

Figure 1.10   Models for molecular architecture of membrane

Lamellar model   According to this model the membrane has a leaflike architecture. The lipid bilayer is formed by the assembly of monolayer in a **"tail to tail"** fashion to form a leaflike structure. The bilayer is sandwiched by a protein layer. The protein layer is attached to **sialic acid** residues.

Micelle model   Micelle model explains lipid molecules as **spherical** structures. Lipid molecules are arranged with their head facing outward and tail facing inward to form a spherical structure with central hydrophobic **core** and an outer hydrophilic **shell** thereby forming a micelle structure. Micelles are covered by protein monolayer.

Protein crystal model   Phospholipid bilayer fills the cavity between protein molecules giving a crystal-like appearance.

Fluid mosaic model   Fluid mosaic model is the widely accepted model for membrane architecture. This model was proposed by Sanger and Nicholson in 1972. According to them membranes have mosaiclike structure made up of both proteins and lipids. The surface of the lipid bilayer has peripheral proteins. Integral proteins are tightly bound to the lipid bilayer. Lipids keep moving and have fluidity. It has more mobility in the fluid state. When the temperature is decreased, lipid changes from fluid state to solid state and shows less mobility. This process is called **phase transition**. During phase transition the carbon–carbon bond in the lipid rotates by $120°$ and assumes a new conformation called **"Gauche"** conformation. The exact temperature at which phase transition occurs is called transition temperature. Lipids with saturated fatty acids are tightly packed and they show less mobility or fluidity. Lipids with unsaturated fatty acids are loosely packed and they show more mobility or fluidity. Also the transition temperature is less for lipids with unsaturated fatty acids.

Cytoskeleton

Cytoskeleton refers to an intricate network of cytoplasmic proteins used to maintain the shape and movement of cell.

Types of cytoskeleton There are three types of cytoskeleton:

1. Microfilaments
2. Intermediate filaments
3. Microtubules

Microfilaments Microfilaments are the **thinnest** and the most **stable** components of the cytoskeleton. They are made up of proteins called **actin**. Actin is a **globular** protein with a head and a tail. Unpolymerized form of actin is called **G-actin**. A number of actin molecules polymerize in a head-to-tail fashion to form a **filamentous** structure of 7 nm diameter called microfilament. Filamentous actin is called **F-actin**. Actin microfilament along with another protein called **myosin** is involved in muscle contraction. Polymerization of actin and formation of microfilament is inhibited by a compound called cytochalasin-B. Formation of microfilament and its function is summarized in Figure 1.11.

Intermediate filaments Intermediate filaments are a rodlike molecular assembly of proteins. They are made up of a diverse group of proteins. The following five types of proteins assemble to form intermediate filaments of approximately 10 nm diameter:

1. Vimentin
2. Desmin
3. Neurofilaments
4. Glial fibrillary acidic protein
5. Cytokeratin

Intermediate filaments act as the **structural component** of cells. It helps in cell–cell attachment, stabilization of epithelial cells, maintenance of the cell structure and providing strength and rigidity to certain cells like nerve cells.

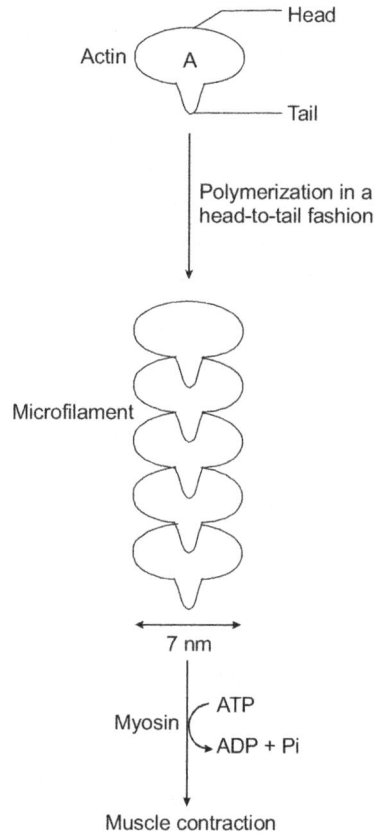

Figure 1.11   Microfilament formation

Microtubules   Microtubules are built from proteins called **tubulin**. Tubulin is a dimer and it has two subunits —

α subunit and β subunit. These subunits assemble in a head-to-tail fashion forming a primary tubular structure called **protofilament**.

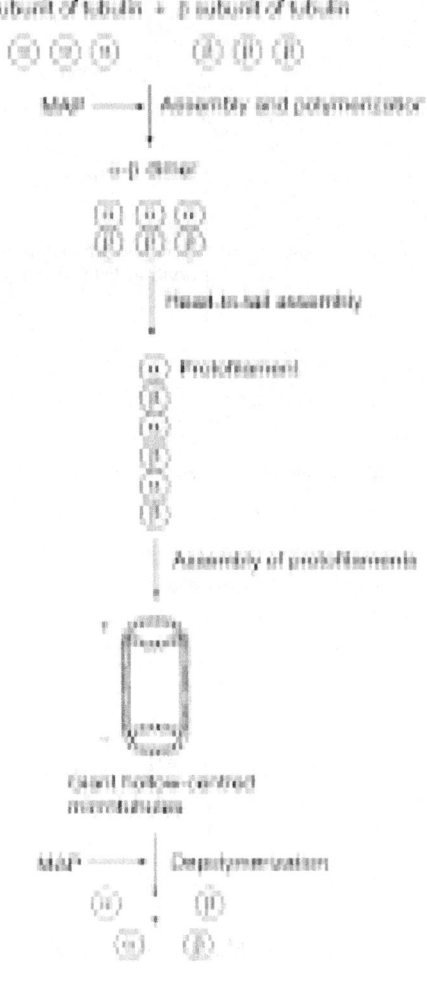

Figure 1.12 Formation of microtubules (MAP—Microtubule-associated protein)

Protofilaments again assemble to form a giant hollow-centred cylindrical structure called microtubules. Microtubule-associated protein (MAP) is responsible for regulating the formation of microtubules. It regulates the polymerization of tubulin subunits (formation of protofilaments) and depolymerization of tubulin subunits (dissociation of protofilaments). Figure 1.12 illustrates the formation and regulation of microtubules.

Functions of microtubules Microtubules are responsible for:

- intracellular and extracellular movement of cell
- movement of cilia and flagella
- movement of macromolecules in the form of vesicles (because of its hollow-centred structure). This is aided by ATP and special proteins called molecular motors (Figure 1.13).

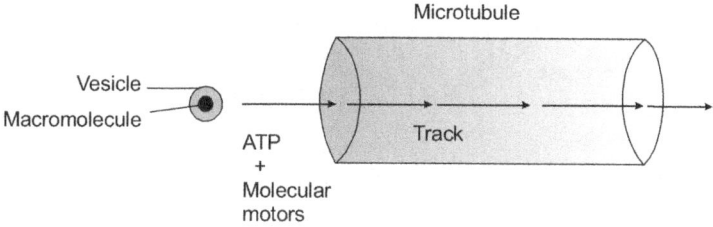

Figure 1.13 Movement of macromolecules through microtubules

CELL DIVISION

Cell division is a process by which the cell duplicates itself for growth and reproduction. Cell division results in two genetically identical daughter cells (Figure 1.14). It is a complex process and includes four phases. These phases are collectively called cell cycle (Figure 1.15).

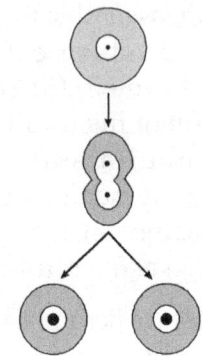

Figure 1.14   Simplified representation of cell division

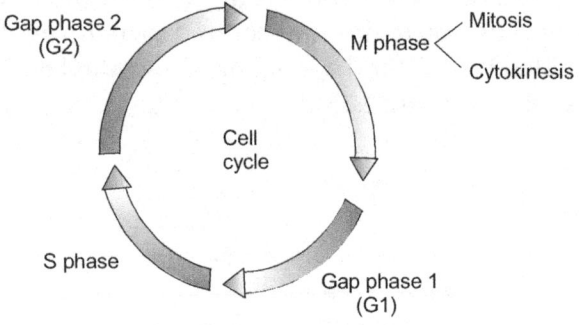

Figure 1.15   Cell cycle

The four phases are:

1. G1 phase
2. S phase
3. G2 phase
4. M phase

G1, S and G2 are collectively called **interphase**.

S phase and M phase are the major phases separated by gap phases.

G1 PHASE—THE FIRST GAP PHASE

Gap phase represents the time taken by the cell to monitor the internal and external environment and the time taken to ensure that the environment is suitable for cell division. During G1 phase, enzymes and substrates necessary for synthesis of DNA (replication) are synthesized and organized.

S PHASE—THE SYNTHETIC PHASE

S phase is the **longest phase** in the cell cycle. In a typical mammalian cell, half of the cell cycle time is spanned by S phase (10–12 hrs). During S phase DNA is synthesized and chromosome is duplicated.

G2 PHASE—THE SECOND GAP PHASE

G2 phase is called **post synthesis** phase in which growth of cytoplasm, formation of organelles and formation of microtubules takes place.

M PHASE—THE MITOTIC PHASE

M phase has two events

1. Mitosis (chromosome segregation and nuclear division).
2. Cytokinesis (cytoplasmic division).

Mitosis

Mitosis is a long process which occurs in five phases: prophase, prometaphase, metaphase, anaphase and telophase (Figure 1.16) shows the various phases in mitosis.

Prophase Nuclear events and cytoplasmic events takes place during prophase. In the nucleus, chromosomes coil and condense and become a rigid structure called "**sister chromatids**". Sister chromatids are attached with one another by means of a protein called **cohesin**.

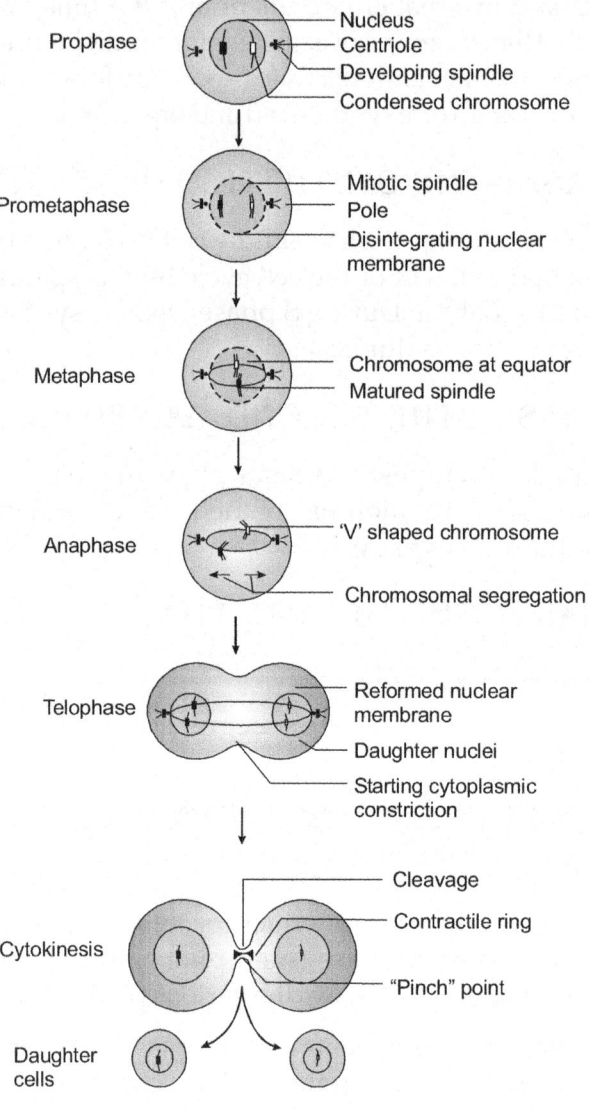

Figure 1.16   Phases in mitosis

In the cytoplasm, on either side of the nucleus, a pair of cylindrical structures with L-shaped configuration are formed at right angles to each other. This is called **centriole**. Centriole elongates and reaches the opposite side of the nucleus, forming the **pole** of the cell. **Microtubules** align themselves forming a giant structure called spindle. From the pole, starlike fibres radiate outwards to form a structure called **aster**.

Prometaphase   This is the **shortest** phase of mitosis. In this phase the nuclear envelope disintegrates and disappears.

Metaphase   Spindle formation is completed in this stage. Spindles extend towards the pole and organize along the equator. Spindle invades the nucleus. Sister chromatids align in the spindle along equator.

Anaphase   In this phase, chromatids segregate due to their movement towards the opposite pole (due to the pulling force of spindle). Chromatid move towards pole at a speed of 0.2–4 µm/minute and assume **'V' shape**.

Telophase   In this phase, spindles disassemble and the nuclear envelope reassembles. Daughter chromosomes are packaged in new nucleus. At the end of the telophase, cytoplasm gets ready for cytokinesis.

Cytokinesis

Filamentous proteins formed in G2 phase bring about cytokinesis. Actin and myosin form a filamentous structure and get organized into a ring called **contractile ring.** Contractile ring induces a **cleavage furrow** and pinches off the cytoplasm leading to cytokinesis.

MEIOSIS

Meiosis is a type of cell division that takes place in **germ cells**; mitosis takes place in somatic cells. Meiosis is also called **reduction division** in which the number of chromosome per cell is halved. Before meiosis cells have 46 chromosomes represented as $2n$.

Table 1.5   Differences between mitosis and meiosis

Mitosis	Meiosis
Occurs in somatic cells (body cells).	Occurs in germ cells (sperm and ovum).
Prophase is short.	Prophase is a long phase with six substages.
Chromosomes will duplicate into two chromatids; both chromatids will be inherited.	Only one chromosome of the homologous pair will be passed to each daughter cell.
Chromosome splitting takes place in early stage of prophase.	Chromosome splitting takes place in late stage of prophase.
Crossing over (chiasma formation) does not take place.	Crossing over (chiasma formation) takes place.
In metaphase, chromosomes exist as a diad.	In metaphase, chromosomes exist as a tetrad.
In anaphase, chromosomes have single chromatid (monad).	In anaphase, chromosomes have two chromatids (diad).
Chromosomes are long and thin.	Chromosomes are short and thick.
Telophase always takes place.	Telophase may or may not occur.
Chromosome number in daughter cell is equal to chromosome number in parent cell.	Chromosome number in daughter cell is half of that in parent cell.

During meiosis I the paired homologous chromosomes are separated into two cells resulting in two haploid cells with 23 chromosomes represented as *n*. Thus meiosis I is referred to as reductional division. Meiosis II is similar to mitosis.

Differences between mitosis and meiosis are shown in Table 1.5.

CONTROL OF CELL CYCLE (REGULATION OF CELL CYCLE)

Checkpoints

Cell cycle can be controlled at three points called as checkpoints or regulatory points (Figure 1.17).

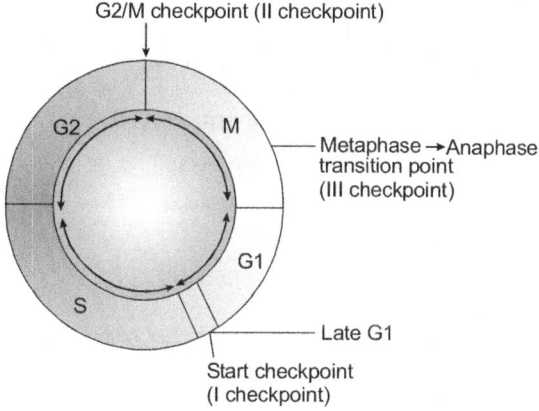

Figure 1.17   Checkpoints for the control of cell cycle

They are:

I checkpoint or late G1 checkpoint

II checkpoint of G2/M checkpoint

III checkpoint or metaphase anaphase–transition point.

At these checkpoints triggering or stopping of cell cycle takes place.

Components in Cell Cycle Regulation

Cyclin (G1/S cyclin, S cyclin, M cyclin) and cyclin-dependent kinase (CdK) are the two components involved in controlling cell cycle at checkpoints.

Cyclin   G1/S cyclin controls the entry into cell cycle.

S cyclin controls mitotic events and synthetic events.

M cyclin triggers G2/M transition.

CdK   Free CdK is inactive and will not exert any control over cell cycle. CdK has a T-loop structure and a deep groove or cleft which is normally closed in free state.

Binding of cyclin to the T-loop exposes the active site and forms active cyclin–CdK complex (Figure 1.18). Cyclin–CDK complex controls cell cycle.

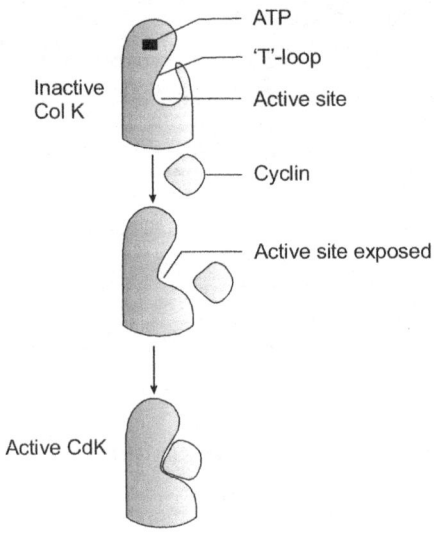

Figure 1.18   Activation of cyclin-dependent kinase (CdK)

Anaphase-promoting complex (APC) APC regulates cell cycle by acting on the 3rd checkpoint – the metaphase-anaphase transition point. It is also known as **cyclosome**. It promotes cell cycle by promoting anaphase. The mechanism of cell cycle progression involves degradation of proteins like **securin** and **cyclin** by a process called ubiquitinylation.

Securin protects the protein linkages that hold the sister chromatids during early mitosis (Figure 1.19). APC induces the addition of ubiquitin to securin. Generally ubiquitinylation signals that the proteins are susceptible to degradation.

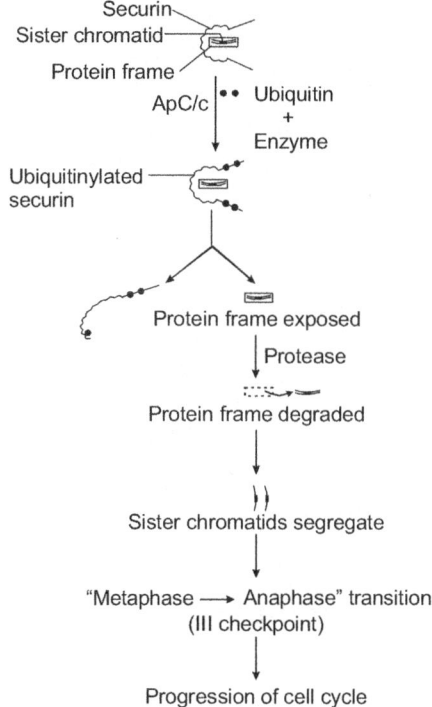

Figure 1.19 Events leading to "metaphase → anaphase" transition by securin

Thus, securin is degraded. Then the protein framework is degraded and the sister chromatids get segregated leading to metaphase–anaphase transition. Consequently cell cycle will progress.

SCF protein   SCF protein is a trimer with three subunits – S, C and F. Activation of cell cycle by SCF protein is shown in Figure 1.20. The F subunit has its structure complementary to another protein called F-box protein. Thus F subunit aids the attachment of SCF protein to F-box protein.

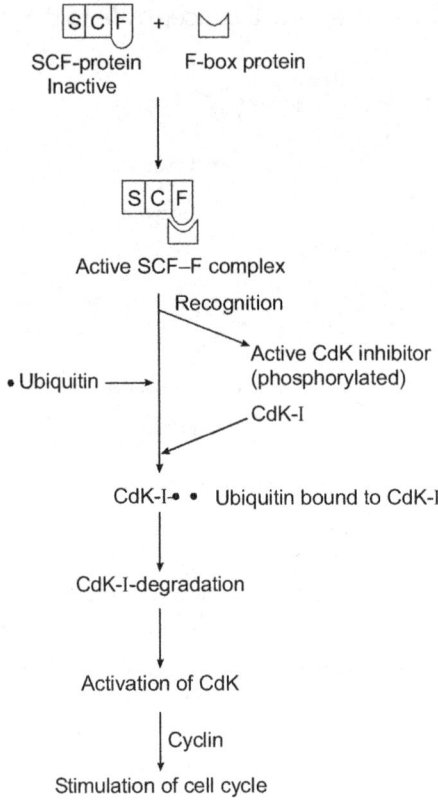

Figure 1.20   Cell cycle stimulation by SCF protein

Thus an active SCF–F box complex is formed. This complex recognizes the active form of CdK inhibitor, CdKI (phosphorylated form), and adds ubiquitin to it. Ubiquitin–conjugated CdKI is recognized by protease enzyme and degraded by cleavage of peptide bond. This results in degradation of CdKI and activation of CdK. Activated CdK in turn binds to cyclin and stimulates cell cycle.

Transcriptional control Transcriptional control refers to control of cell cycle at gene level. Two types of genes are involved in cell cycle control. Promoter gene (P) and suppressor/inhibitor gene (S/I). When P gene is switched "on" proteins that promote cell cycle are translated. When S/I gene is switched "on" proteins that suppress cell cycle are produced (Figure 1.21).

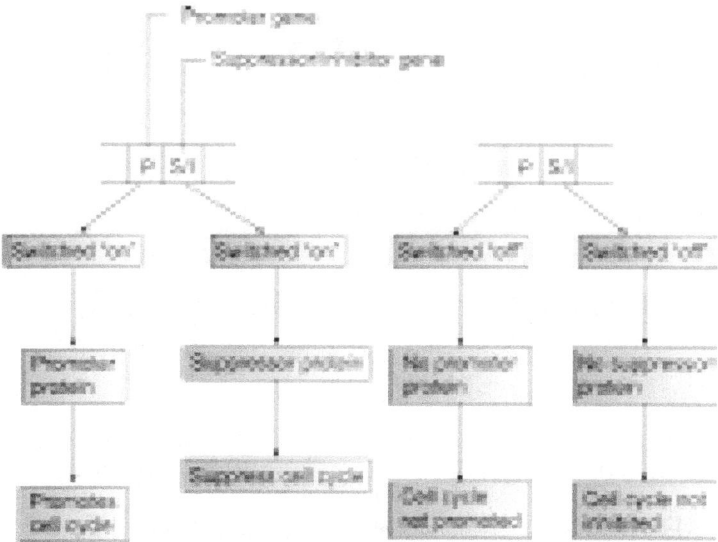

Figure 1.21   Cell cycle control at transcriptional level

FUNGI

Fungi are spore-forming eukaryotes.

STRUCTURE

Body of fungi is called thallus (plural: thalli). It has a cell wall which is strong, and made up of nitrogen-containing polysaccharides. Mostly, fungal cell wall is made up of polymers of N-acetyl glucosamine and is called chitin. Thallus is the vegetative structure of fungi. Depending on the nature of the thallus, fungi can be classified into four types:

1. *Yeast form (unicellular fungi)*   Cells occur singly and one end of the thallus bulges to form polar bud. It has chitin, plasma membrane, nucleus and mitochondria (Figure 1.22a), e.g. *Saccharomyces cerevisiae.*

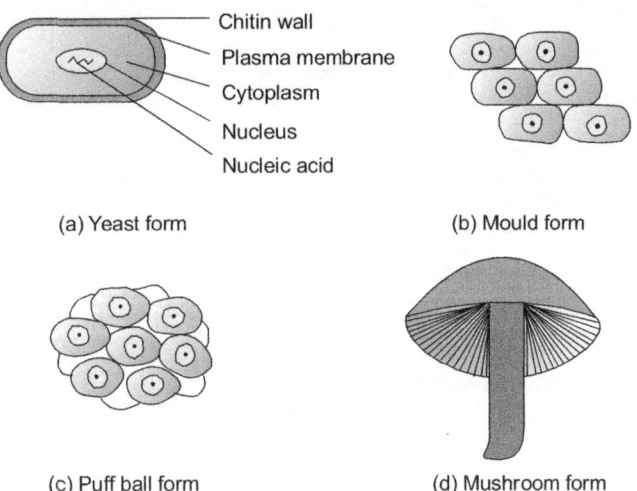

Chitin wall
Plasma membrane
Cytoplasm
Nucleus
Nucleic acid

(a) Yeast form   (b) Mould form

(c) Puff ball form   (d) Mushroom form

Figure 1.22   Thallus structure of different forms of fungi

2. *Mould form (multicellular fungi)*   These are the multicellular forms. Cells are long, threadlike, branched filamentous structures which form a

network called **hyphae** (Figure 1.22b), e.g. *Penicillium* species. Many hyphae aggregate to form a tissue called **mycelium.**

3. *Puff ball form (macroscopic fungi)* Cells aggregate into clusters with puffed ball structure (Figure 1.22c), e.g. *Lycoperdon* species.

4. *Mushroom form (macroscopic fungi)* Thallus is very well developed and densely packed with an umbrella like appearance. It has a cap and a stem. Under the cap are fine threadlike structures called gills (Figure 1.22d), e.g. *Gyromitra* species.

FUNGAL DIMORPHISM

Dimorphism is the capacity of pathogenic (disease-causing) fungi to exist in two forms, say, yeast form (Y) and mould (M) form. This is also called Y ↔ M shift or Y ↔ M dimorphism. Fungal dimorphism can simply be represented as:

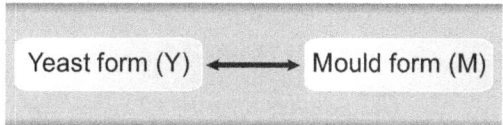

Yeast form (Y) ←——→ Mould form (M)

Y ↔ M Shift

Animal fungi exist in Y form. When exposed to certain environmental conditions like nutrient deficiency, CO_2 tension, temperature change, redox potential, etc. they get transformed to M form. This is called Y ↔ M shift. It is represented as follows:

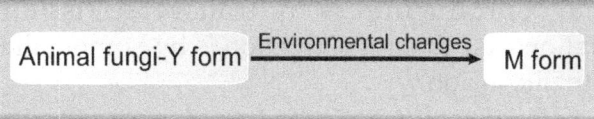

Animal fungi-Y form ——Environmental changes——→ M form

M ↔ Y Shift

Plant fungi exist in M form. When exposed to certain environmental conditions like nutrient deficiency, CO_2 tension, temperature change, redox potential, etc. they get transformed to Y form. This is called M ↔ Y shift. It is represented as follows:

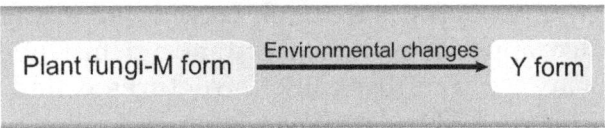

Plant fungi-M form   Environmental changes ⟶ Y form

Some examples for fungi exhibiting dimorphism include *Blastomyces dermatitidis, Candida albicans, Coccidiodes immitis, Histoplasma capsulatum, Sporothrix schenckii* and *Paracoccidiodes brasiliensis.*

FUNGI EXHIBITING DIMORPHISM
Blastomyces dermatitidis
Candida albicans
Coccidiodes immitis
Histoplasma capsulatum
Sporothrix schenckii
Paracoccidiodes brasiliensis

REPRODUCTION IN FUNGI

Reproduction in fungi is by asexual and sexual methods; Asexual reproduction is by transverse fission and sporulation. Asexual reproduction is by homothallic and heterothallic fusion.

TYPES OF REPRODUCTION IN FUNGI	
Asexual reproduction	**Sexual reproduction**
Transverse fission	Homothallic fusion
Sporulation	Heterothallic fusion

Asexual Reproduction

Transverse fission   Cells undergo mitosis, followed by cytoplasmic constriction and cell division resulting in the formation of two daughter cells.

Sporulation   Fungal hyphae undergo cell wall splitting and form small fragments from their tip. These fragments are called spores which are capable of developing into new individual cells. Different types of spores are formed during sporulation (Figure 1.23). They are:

NAME OF THE SPORE FORMED DURING SPORULATION
Arthospore
Chlamydospore
Sporangiospore
Conidiospore
Blastospore

- *Arthospore*   Spores fragmented due to splitting of cell wall.
- *Chlamydospore*   Each cell in hypha is surrounded by new cell wall and then separated as chlamydospore.
- *Sporangiospore*   A sac develops at the hyphal tip. It is called sporangium. Spores develop within the sporangium and are liberated out as sporangiospores. Hyphal tip bearing the sporangium sac is called sporangiophore.

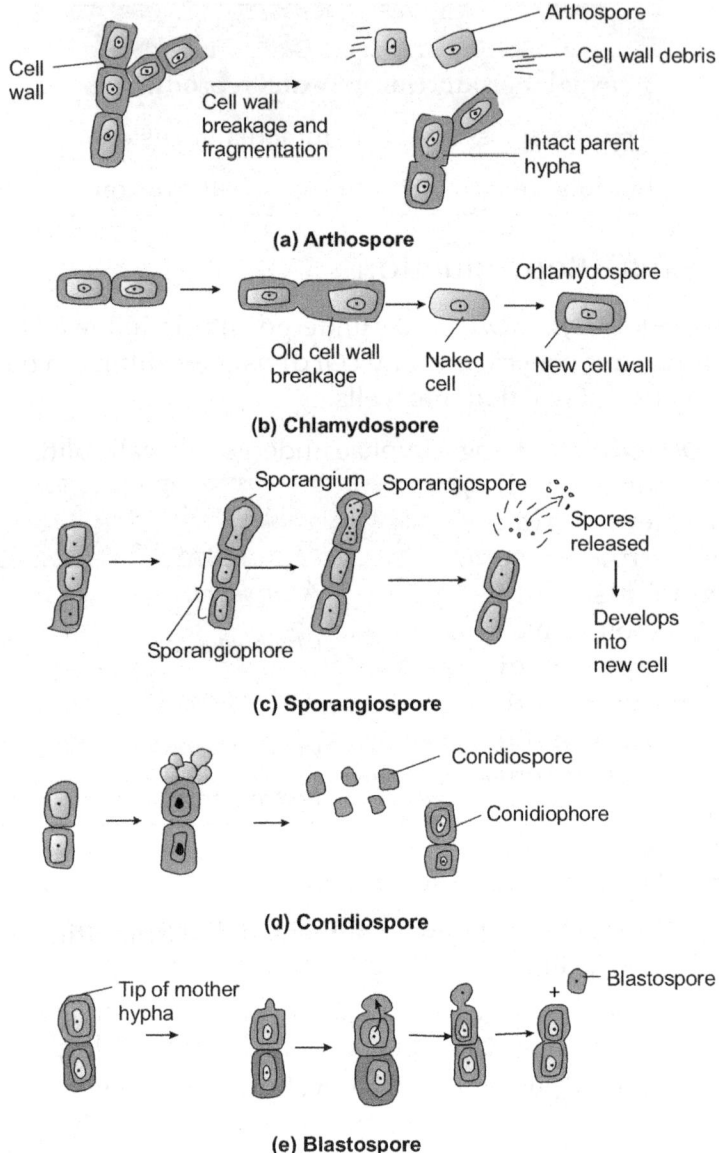

Figure 1.23 Spores formed during sporulation in fungi

- *Conidiospore*   Spores develop in the hyphal tip in the absence of sac and liberated as conidiospore. Hyphae bearing the spores are called conidiophores.
- *Blastospore*   Spores are produced by budding of mother cell.

Sexual Reproduction

Homothallic fusion   Fusion between two compatible nuclei that exist in same thallus or same mycelium is called homothallic fusion (Figure 1.24a).

Heterothallic fusion   Fusion between two compatible nuclei that exist in different thalli or different mycelia is called heterothallic fusion (Figure 1.24b).

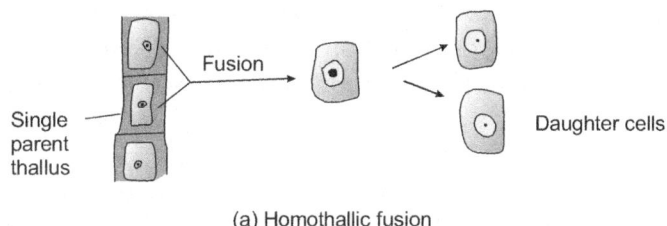

(a) Homothallic fusion

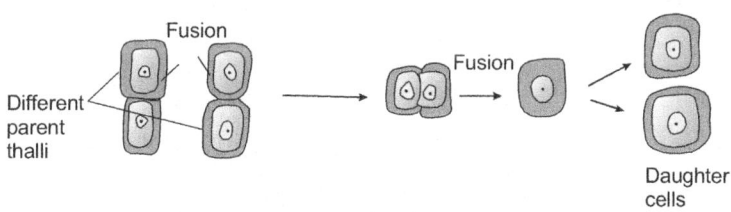

(b) Heterothallic fusion

Figure 1.24 Homothallic and heterothallic fusion in fungi (Sexual reproduction)

ALGAE

Algae are a diverse group of photosynthetic eukaryotic organisms (protists). They survive in moist areas independently (solitary algae) or in association with fungi (lichens).

STRUCTURE

Structure of algal cell is similar to that of multicellular organisms (Figure 1.25). In addition to the common cell organelles, they possess chloroplast, pyrenoid and pellicle. The outer cell membrane is called plasmalemma. Cytoplasm is divided into outer region called ectoplasm which gives rigidity and stability to the cell and inner area called endoplasm which accommodates the cell organelles. Two types of vacuoles are present in the endoplasm: phagocytic vacuoles used in food digestion and contractile vacuole used in osmoregulation. The outer membrane is closely embedded on the ectoplasm and both are together called pellicle. Chloroplasts used for photosynthesis are found in the cytoplasm. Chloroplast is made up of dense region in the centre which is called pyrenoid. Pyrenoid is used to store starch.

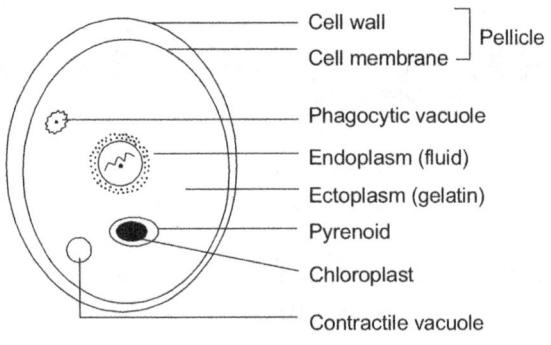

Figure 1.25   Algal cell

ENCYSTMENT AND EXCYSTMENT

Algae undergo two phenomena called encystment and excystment. Cells become dormant showing low metabolic activity and attain a resting stage. This is called encystment. Such a dormant structure is called cyst. Cyst protects the cell against adverse environmental conditions like nutrient deficiency, desiccation, adverse pH and low oxygen.

Algae become active under favourable conditions and skip cyst formation. This phenomenon is called excystment.

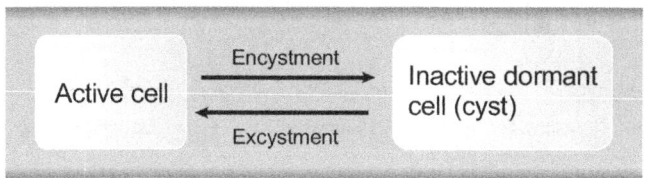

REPRODUCTION IN ALGAE

Algae reproduce by asexual reproductive modes like binary fission, multiple fission and budding. Sexual modes of reproduction like conjugation and syngamy are also possible. In binary fission nucleus assumes dumbbell shape and undergoes mitosis, and cytoplasm undergo cytokinesis. This results in the division of parent cell into two identical daughter cells (Figure 1.26a). In multiple fission multiple copies of daughter cells are formed after cell division. In budding the polar end of the cell becomes active and forms a projection which develops into bud. Nuclear material gets transferred into the bud and the bud gets detached from the parent and grows into new cell (Figure 1.26b).

In sexual reproduction gametes are involved.

During conjugation, two different cells mate and exchange their nuclear content and fuse into a single cell. This hybrid cell then redivides into two new daughter cells.

MODES OF REPRODUCTION IN ALGAE	
Asexual reproduction	**Sexual reproduction**
Binary fission	Conjugation
Multiple fission	Syngamy
Budding	

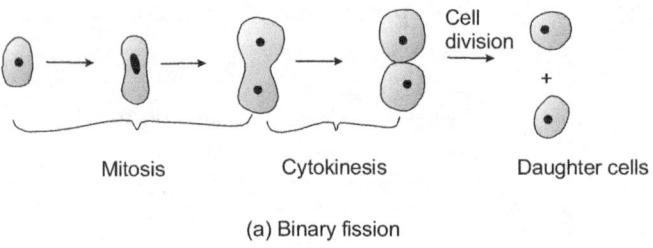

(a) Binary fission

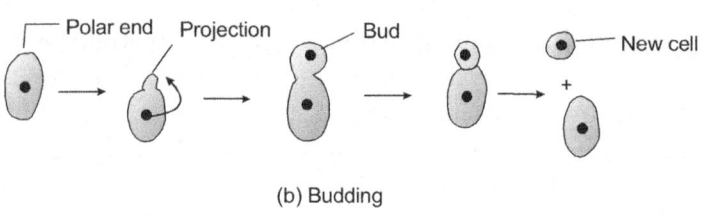

(b) Budding

Figure 1.26 Asexual reproduction in algae

Syngamy refers to the process of fusion. Each cell functions as a gamete. Two gametes fuse to form a zygote which develops into new cell. Two types of syngamy are seen in algae:

- *Isogamy* Fusion of two morphologically similar gametes.
- *Anisogamy* Fusion of two morphologically dissimilar gametes.

PROKARYOTES
BACTERIAL CELL

SHAPE AND SIZE

Bacteria (singular: Bacterium) are unicellular prokaryotic microorganisms. They exist in different shapes and are named accordingly (Table 1.6). For example, spherical cells are called cocci (singular: coccus), e.g. *Streptococcus pyogenes, Streptococcus mutans, Lactococcus lactis*. Rod-shaped bacterial cells are called bacilli (Singular: bacillus); e.g. *Bacillus megaterium*. Bacteria with comma-shaped cells or curved cells are called vibrios, e.g. *Vibrio cholerae*. Spiral bacteria are called spirilla (Singular: spirillum), e.g. *Rhodospirillum rubrum*. Size of bacterial cells ranges from 1.1–1.5 μm width and 2–6 μm length.

Table 1.6 Nomenclature of bacteria based on the shape of the cells

Shape of the cell	Name	Example
Spherical	Coccus	*Streptococcus pyogenes*
Rod	Bacillus	*Bacillus subtilis*
Comma (curve)	Vibrio	*Vibrio cholerae*
Spiral	Spirilla	*Rhodospirillum rubrum*

STRUCTURE

Ultrastructure of bacteria is shown in Figure 1.27. Various structural components of bacterial cell and their functions are summarized in Table 1.7.

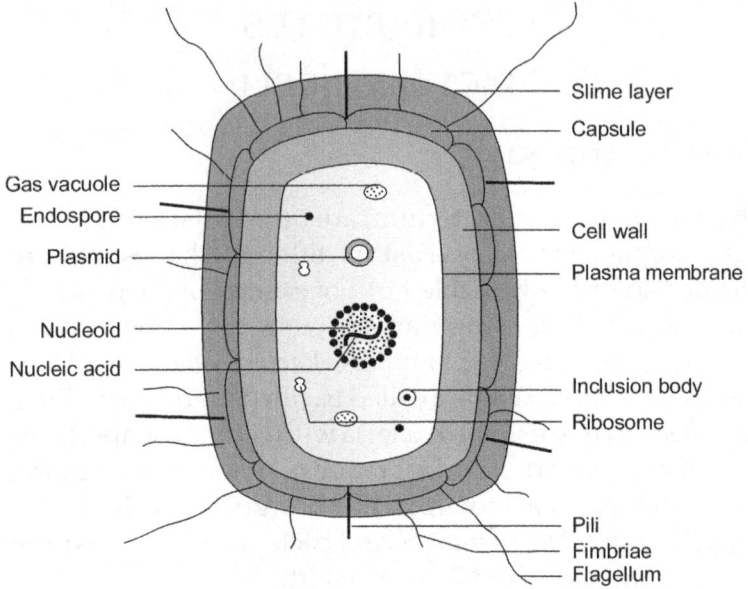

Figure 1.27   Ultrastructure of bacterial cell

Structures Forming the Surface and Internal Part of the Cell

The cell is enclosed by an outer wall called **cell wall**. It gives shape to the cell and protects the cell from osmotic pressure. Internal to the cell wall is the cell membrane also called **plasma membrane** made up of lipid molecules called **hopanoids**. It serves as a mechanical barrier, site for respiration, photosynthesis, and chemotaxis and aids transport of materials across the cell. The inner layer of the membrane is folded forming vesicles and foldings called **mesosomes**. Mesosomes vary in shape based on which they are called spherical, lamellar (leaflike), tubular and flat mesosomes (Figure 1.28). Between the cell wall and the plasma membrane is a space called **periplasmic space** which stores hydrolytic enzymes. Inside the cell is the large area called cytoplasm.

Table 1.7 Bacterial cell structures and functions

Structure	Function
Structures on surface and in inner part of the cell	
Cell wall	Gives shape to cell
	Offers protection from osmotic pressure
Plasma membrane	Mechanical barrier
	Site for transport, respiration, photosynthesis
Periplasmic space	Stores enzymes for nutrient processing
DNA	Genetic material
Gas vacuole	Buoyancy for floating
Ribosomes	Protein synthesis
Inclusion bodies	Storage site
Structures outside the cell wall	
Capsule and slime layer	Cell adhesion
	Resistance to phagocytosis
Fimbriae	Cell adherence
Pili	Reproduction
Flagella	Locomotion

The centre of the cytoplasm has a discrete region called **nucleoid**, where the **DNA** is stored. Various structures are found surrounding the nucleoid. Most important is the extrachromosomal circular DNA called **plasmid or cytoplasmic DNA**. Ribosomes which are needed for protein synthesis are scattered in the cytoplasm. Few vacuoles are found in the cytoplasm. These are called **gas vacuoles** and they provide **buoyancy** when cells float. Some small vesiclelike structures are also found in the bacterial cell. These are called **inclusion bodies.** They act as storage vesicles in which various compounds like sodium, potassium and starch are stored. Endospores are small spherical structures which offer survival capacity to the cells under harsh environmental conditions.

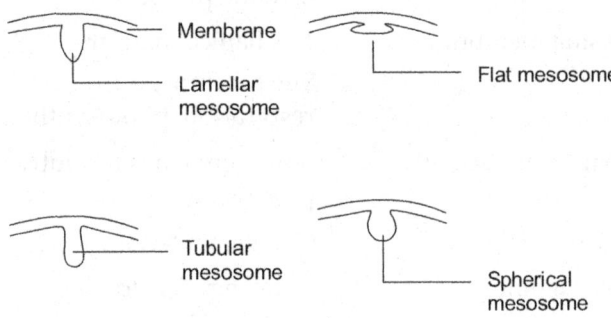

Figure 1.28   Mesosomes in bacterial cell

Magnetosomes   Magnetosomes are specialized inclusion bodies found in the cytoplasm of magnetotactic bacteria and helps orienting the cells in earth's magnetic field. *Aquaspirillum magnetotacticum* is an example for bacteria possessing magnetosomes. The structure of magnetosome is shown in Figure 1.29. Magnetosomes are 35–125 nm diameter. They have an outer lipid layer coat and an inner matrix. Embedded in the matrix is the chain of iron spheres. Each sphere functions as a tiny magnet. Iron exists in three forms:

1. Magnetite (Fe_3O_4)
2. Gregite (Fe_3S_4)
3. Pyrite (FeS_2)

When bacteria are exposed to magnetic field, magnetosomes help to give a wavelike motion.

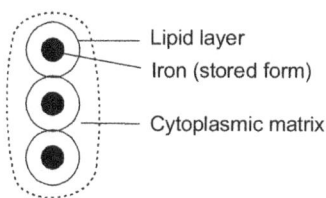

Figure 1.29   Magnetosome

Structures Forming the Outer Part of the Cell

Bacteria have some structures attached to the external surface of the cell wall. These include:

- Outer layer (capsule/slime layer/glycocalyx).
- Fimbriae
- Pili
- Flagella

Outer layer   Outer layer is present outside the cell wall. If the layer is strong and rigid and well organized it is called **capsule**. Capsule cannot be washed off with chemicals. If the layer is unorganized forming a diffused zone, it is called **slime layer.** It can easily be washed off with chemicals. If the outer layer is formed of interconnected network of polysaccharides it is called **glycocalyx**. These layers are used for adherence or attachment of bacteria to their substrate and also help in offering resistance to phagocytosis.

Fimbriae   Fimbriae is a very small slender tubular structure 3–10 nm in diameter. They are made up of helically arranged proteins (proteins folded into secondary structure) (Figure 1.30).

Each bacterium has approximately 1000 fimbriae/cell. Fimbriae function in **cell attachment** and **cell movements** like gliding movement and jerking movement.

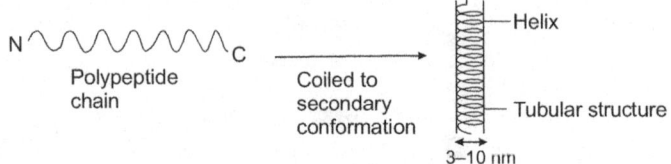

Figure 1.30   Fimbriae

Pili   Pili are a slender hairlike structures larger than fimbriae. They are 9–10 nm in diameter. They have a wavy appearance. Pili are responsible for conjugation (reproduction) and attachment of virus to the cell surface (Figure 1.31).

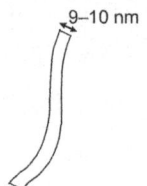

Figure 1.31   Pili

Flagella   Flagellum (singular) is a threadlike thin outgrowth. It is rigid and is about 15–20 μm in length and 20 μm diameter. It is used for locomotion. The structure of the flagellum is shown in (Figure 1.32). It emerges from the deep interior of the cell through the basal body. The body of the flagellum is called flagellar filament which is made up of protein subunits called **flagellin**. Flagellar filament is attached to the basal body via the **flagellar hook**. The tip of the filament has a protein called **capping protein**. Based on the number of flagella and location of flagella, bacteria can be called by different names (Table 1.8). Figure 1.33 shows different types of bacteria classified based on flagellar arrangement. Bacteria with one flagellum at one end are called **monotrichous** and those with flagella at both ends are called **amphitrichous**. Bacteria with a tuft of flagella at one end are called **lophotrichous**. Bacteria with flagella distributed throughout the outer surface of the cell are called **peritrichous**.

Table 1.8   Types of bacteria based on flagellar arrangement

Number of flagella	Location	Nomenclature
One	One side of the cell	Monotrichous
Two	One at each side of the cell	Amphitrichous
Many	One side of the cell (Tuft)	Lophotrichous
Many	Throughout the surface of the cell	Peritrichous

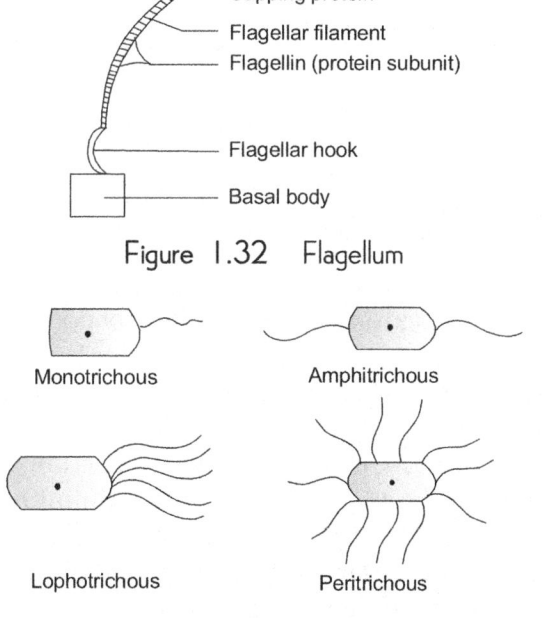

Figure 1.32   Flagellum

Figure 1.33   Types of bacteria based on flagella

GRAM-POSITIVE AND GRAM-NEGATIVE BACTERIA

Bacteria are classified as gram-positive and gram-negative based on the differences in the structure of cell wall. Differences between gram-positive and gram-negative bacteria are shown in Table 1.9 and diagrammatically represented in Figure 1.34.

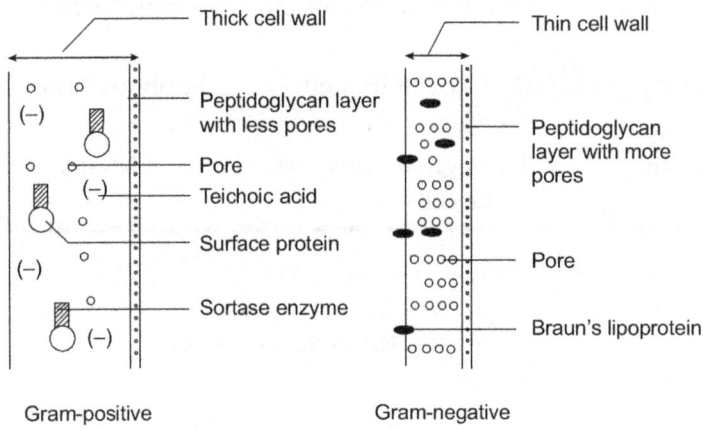

Gram-positive   Gram-negative

Figure 1.34 Structure of cell wall in gram-positive and gram-negative bacteria

One group of bacteria has thick cell wall. The cell wall is highly cross linked and has less number of pores of smaller size. These bacteria are called **gram-positive** bacteria. Other groups of bacteria which have thin cell wall with less cross links and more number of larger pores are called **gram-negative** bacteria. The term "Gram" is used because the differences can be identified using a staining technique discovered by a scientist called Christian Gram. Hence the technique is called Gram staining. In this technique bacteria take up the dye differently and stain differently due to differences in their cell wall structure.

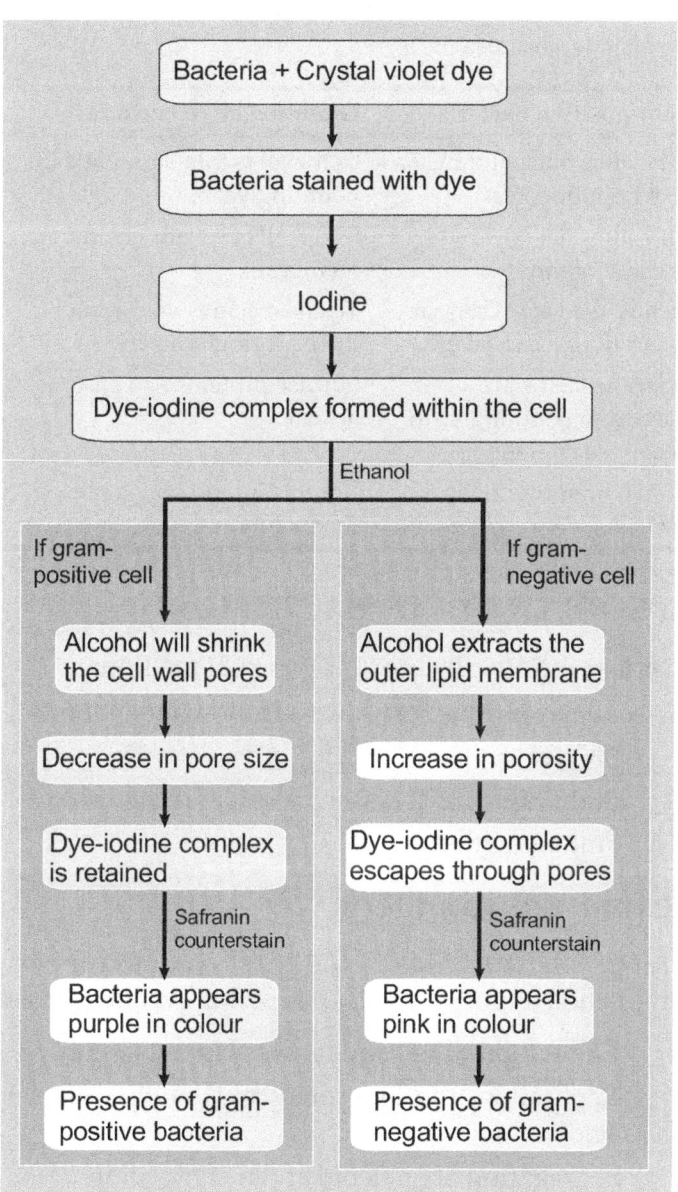

Table 1.9   Differences between gram-positive and gram-negative bacteria

Gram-positive bacteria	Gram-negative bacteria
Cell wall is thick and made up of peptidoglycan.	Cell wall is thin and made up of peptidoglycan.
Cell wall is not covered by outer membrane.	Cell wall is covered by outer membrane.
Teichoic acid is present in the peptidoglycan layer.	Teichoic acid is not present in the peptidoglycan layer.
Surface proteins are attached to peptidoglycan via sortase enzyme.	Surface proteins and sortase are absent.
Braun's lipoprotein is absent.	Braun's lipoprotein is present in the membrane.

REPRODUCTION IN BACTERIA

Bacteria reproduce by asexual or sexual method:

Asexual reproduction is by the following methods

- Binary fission
- Budding
- Fragmentation

Asexual Reproduction

Binary fission   This is the most common mode of reproduction (Figure 1.35). The following events take place.

- Elongation of cell and increase in cell length.
- Replication of chromosomes and doubling of nucleic acids.
- Segregation of nucleoid and nucleic acids.
- Constriction of cytoplasm.

- Formation of septum or cross wall between cytoplasm along the constriction.
- Division of cell into two daughter cells.

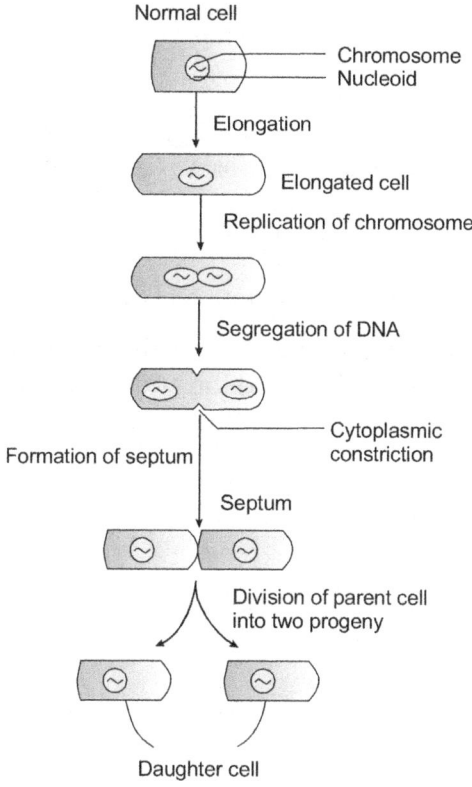

Normal cell — Chromosome / Nucleoid

Elongation

Elongated cell

Replication of chromosome

Segregation of DNA

Formation of septum — Cytoplasmic constriction

Septum

Division of parent cell into two progeny

Daughter cell

Figure 1.35 Binary fission in bacteria

Bacterial reproduction leads to multiplication of bacterial population and this can be represented in the form of growth curve as shown in Figure 1.36. During the initial stage the number of cells remains constant and this phase is called **lag phase**. This is followed by a steady increase in the population which is called **exponential phase**. After a maximal population is reached, the number of cells remains constant and it is called **stationary phase**. Finally the number of cells begins to fall and it is called **death phase** or **decline phase.**

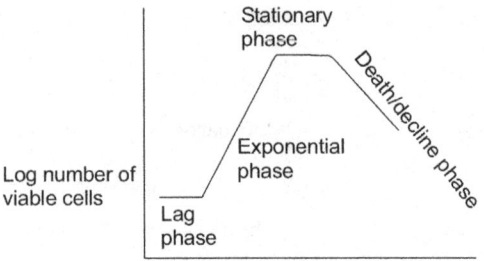

Figure 1.36   Growth curve representing bacterial population

Budding   During this process the tip of the cell is slightly bulged in the form of bud and the bud eventually gets split and pinched off from the cell. Buds are thus separated from the parent and later developed into new cell (Figure 1.37).

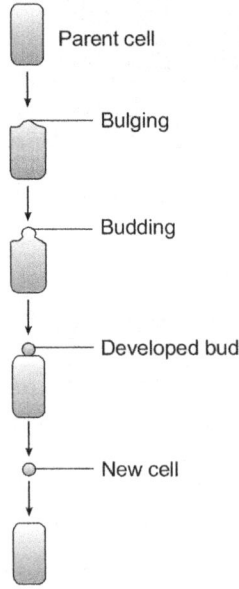

Figure 1.37   Budding mode of reproduction in bacteria

Fragmentation In this mode of reproduction a new organism grows from a fragment of the parent. Each fragment develops into a mature, fully grown individual. This process is common in photosynthetic bacteria.

Reproduction by Exchange of DNA/Sexual Reproduction

Sexual reproduction involves exchange of DNA material between cells. There are three types of sexual reproduction:

1. **Conjugation** in which male cell passes DNA to female cell by means of a conjugation tube (sex pilus).
2. **Transformation** in which bacterium takes up DNA released by dead bacteria.
3. **Transduction** in which bacteriophages carry DNA from one cell to another.

VIRUS

Viruses are highly infective microorganisms, 10–400 nm in size. They cannot survive independently and hence need a host to infect and survive.

STRUCTURE OF VIRUS

A complete virus structure is called "virion" (Figure 1.38). It has three components:

1. Genetic material
2. Capsid
3. Envelope

Genetic material in virus is either DNA and RNA. Based on this virions are called DNA virus or RNA virus. The genetic material of the virus usually gets integrated with the host chromosome.

Capsid is a protein coat which encloses the DNA or RNA. Capsid and the enclosed genetic material can together be called **nucleocapsid**. Proteins in the capsid are assembled in the form of subunits called **protomeres**. Two types of capsid are found in virus: 1) Helical capsid and 2) icosahedral capsid. In the helical capsid, the protein subunits are arranged in a spiral manner having a helical conformation. It is hollow, rigid and tubular. For example, tobacco mosaic virus (TMV) has a helical capsid. In icosahedral capsid the protein subunits are arranged in icosahedral fashion and these subunits are called **capsomeres**. For example, Simian virus-40 (SV40) has icosahedral capsid enclosing the DNA.

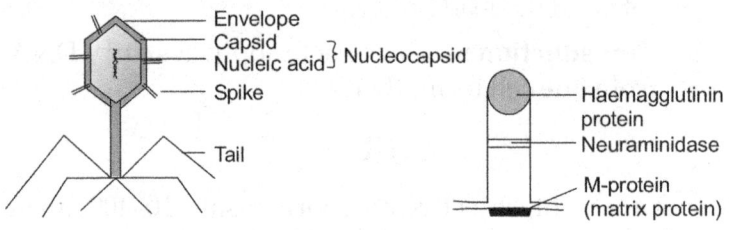

Figure 1.38   Structure of virus   Figure 1.39   Structure of spike

Envelope is a membranous outer covering made up of lipid. They are flexible. They have protein subunits called peptomeres. **Peptomeres** form a projection called **spike** (Figure 1.39), which is useful for attachment to the host. It is 10 nm in length and occurs at a distance of 7–8 nm. It is made up of three components—**matrix protein** at its base which helps stabilizing the spike, **haemagglutinin protein** which forms precipitate with red cells and an enzyme called **neuraminidase** which helps release of mature virions from host cell. Some viruses do not possess envelope and those are called **naked viruses.**

REPRODUCTION IN VIRUS

Viral reproduction involves six steps (Figure 1.40).

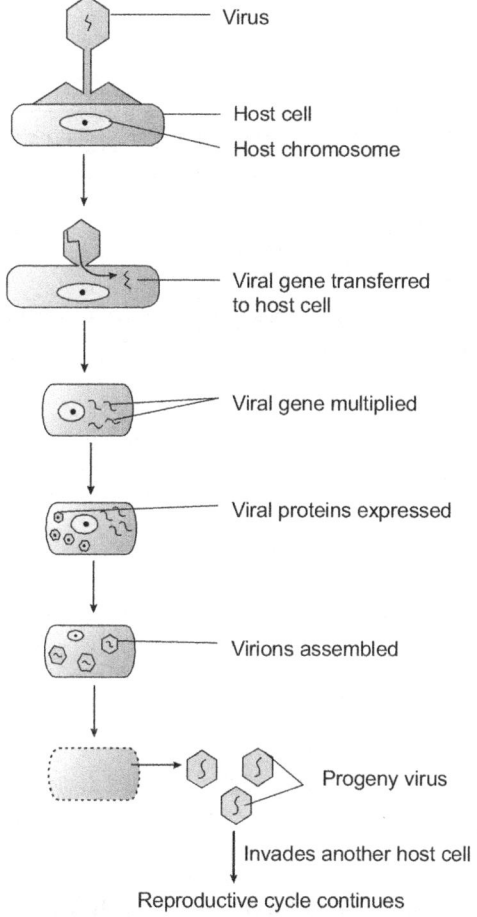

Figure 1.40 Reproduction in virus

1. Attachment of virus to host cell.
2. Entry of nucleocapsid or nucleic acid into the host.
3. Gene expression within the host.

4. Synthesis of viral proteins and nucleic acids.

5. Self assembly of virions.

6. Rupture of host cell and release of virions (progeny virus).

The cycle continues.

REVIEW QUESTIONS

1. Write a note on nucleus.

2. What is endoplasmic reticulum?

3. Differentiate between rough endoplasmic reticulum and smooth endoplasmic reticulum.

4. Define lysosomes and peroxisomes.

5. Write the differences between peripheral and integral membrane proteins.

6. What is gap phase?

7. What is telophase?

8. What is cytokinesis?

9. What is APC?

10. Define mesosomes.

11. Write a note on magnetosomes.

12. Write a brief note on pili and its function.

13. Distinguish between gram-positive and gram-negative bacteria.

14. Explain about encystment and excystment in algae.

CHOOSE THE CORRECT ANSWER

1. Powerhouse of the cell is called
   (a) lysosome   (b) ribosome
   (c) mitochondria   (d) nucleus
2. Vacuole of plant cell is called
   (a) protoplast   (b) centrosome
   (c) tonoplast   (d) chloroplast
3. Dictyosomes are
   (a) Golgi bodies   (b) ribosome
   (c) inclusion bodies (d) vacuoles

SIR ALEXANDER R. TODD
Awarded Nobel Prize for his work on
nucleotides and nucleotide coenzymes.

Nucleic acids are polymers of nucleotides which function
as genetic material.

NUCLEOTIDES

Nucleotides are monomers or building blocks of nucleic
acids which are composed of three structural units.

Pentose Sugar

Two types of pentose sugars are present in nucleic acids.
They are **ribose** and **deoxyribose**. If the pentose is ribose,
the nucleic acid is called **RNA** (ribonucleic acid). If the
pentose is deoxyribose the nucleic acid is called **DNA**
(deoxyribonucleic acid). The structures of these two sugars
are shown in Figure 2.1. Carbon atoms in pentose sugar
are denoted as 1' through 5'.

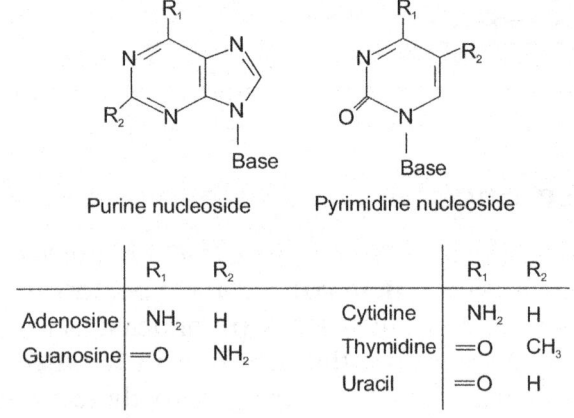

$$5'$$
$$HOCH_2 \quad O \quad OH$$
$$4' \quad \quad 1'$$
$$H \quad H_{3'} \quad 2'H \quad H$$
$$OH \quad H$$

Deoxyribose

$$5'$$
$$HOCH_2 \quad O \quad OH$$
$$4' \quad \quad 1'$$
$$H \quad H_{3'} \quad 2'H \quad H$$
$$OH \quad OH$$

Ribose

Figure 2.1 Structure of sugars found in nucleic acids (Ribose is found in RNA; deoxyribose is found in DNA).

Nitrogenous Base

Two types of nitrogenous bases are found in nucleic acids. They are purine bases and pyrimidine bases. Adenine (A) and guanine (G) are purine bases. Cytosine (C), thymine (T) and uracil (U) are pyrimidine bases. Structure of purine and pyrimidine bases are shown in Figure 2.2. DNA is composed of A, G, C and T. RNA is composed of A, G, C and U. Pentose sugar and nitrogenase base are collectively called as **nucleoside**.

Purine nucleoside   Pyrimidine nucleoside

	R_1	R_2		R_1	R_2
Adenosine	NH_2	H	Cytidine	NH_2	H
Guanosine	$=O$	NH_2	Thymidine	$=O$	CH_3
			Uracil	$=O$	H

Figure 2.2 Structure of purine and pyrimidine bases

Phosphate Group

Phosphate is attached through the oxygen atom of the hydroxyl group of the 5′ carbon of the pentose sugar. Thus a nucleotide can be named as **nucleoside monophosphate.** The structure of a nucleotide with adenine base for example is shown in Figure 2.3.

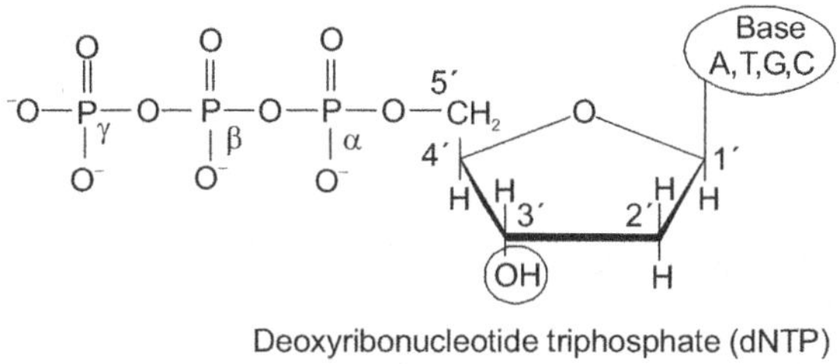

Deoxyribonucleotide triphosphate (dNTP)

Figure 2.3 Structure of a nucleotide

STRUCTURE OF NUCLEIC ACIDS

STRUCTURE OF DNA

The primary, secondary and tertiary structures of DNA are given in Figure 2.4.

Primary Structure of DNA

Primary structure of DNA refers to the **phosphodiester bond backbone.** In nucleic acid strands, nucleotides are linked by phosphodiester bond. This is a bond formed between the 5′-phosphate group of one nucleotide and 3′-hydroxyl group of the adjacent nucleotide through a phosphate molecule. A schematic representation of the phosphodiester bond is shown in Figure 2.4a.

A single nucleic acid strand formed by phosphodiester bond has two termini namely 5′ end with phosphate group and 3′ end with hydroxyl group generally represented as P-5′ ----------- 3′-OH. Conventionally base sequences of DNA are written in the 5′ → 3′ direction.

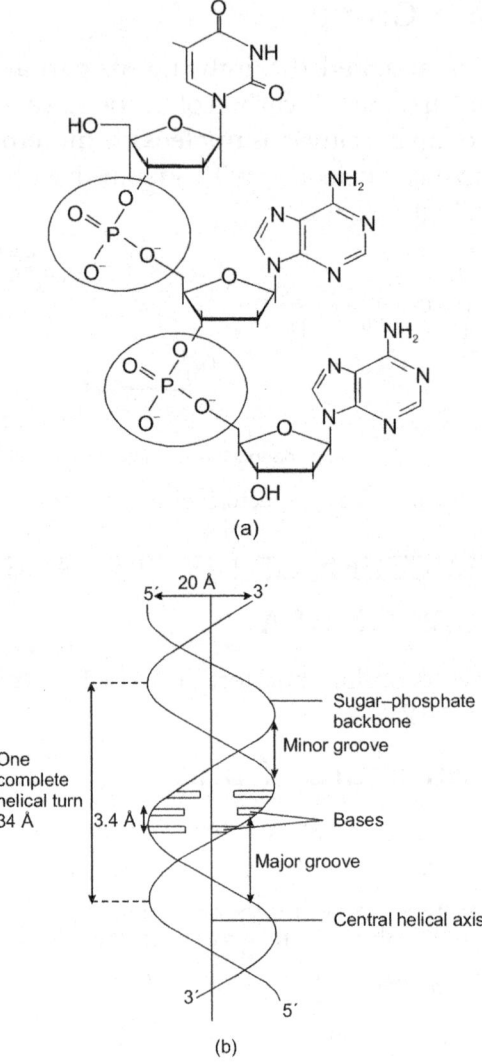

(a)

(b)

Figure 2.4   Structure of DNA (a) Primary structure—phosphodiester bond backbone (b) Secondary structure—double helix structure (c) Tertiary structure—supercoiled structure (*Continues*)

(c)

Figure 2.4   Structure of DNA (a) Primary structure—
phosphodiester bond backbone (b) Secondary
structure—double helix structure (c) Tertiary
structure—supercoiled structure

Secondary Structure of DNA

Secondary structure of DNA studied using X-ray diffraction
data is called double helical DNA. Double helical DNA
model was proposed by Watson and Crick. This model is
also called B-form DNA (Figure 2.4b). DNA double helix is
made up of two DNA strands running in opposite direction,
i.e., in **antiparallel fashion.** This means that the 3′ and 5′
end of two strands are in opposite direction and are coiled
to form a right-handed helix. The helix has a common axis.
The two strands of the helix are held together by
complementary base-pairing, i.e., A pairs with T via two
hydrogen bonds and G pairs with C via three hydrogen
bonds. Therefore, in a double helix, the number of A=T and
number of G=C are equal. In other words, DNA will have
1:1 ratio of purine and pyrimidine bases. This is called

Chargaff's rule. Base pairs lie inside the helix and the sugar phosphate backbone forms the outer part of the helix. Double helix is stabilized by hydrogen bonding and van der Waals interactions of the stacked base pairs. Double helix is 20 Å wide and 34 Å long. It has a major groove 22 Å wide and a minor groove 12 Å wide.

Tertiary Structure of DNA

Supercoiled DNA, chromosomes and chromatin are the three types of tertiary structure of DNA. DNA can be twisted and coiled into a compact tertiary structure called supercoiled structure which resembles a coiled telephone wire (Figure 2.4c). Supercoiling may be right-handed or left-handed. B-form of DNA is shortened 10,000-fold and folded densely into chromosomes. This occurs during mitosis. DNA also folds into less dense form called chromatin during the interphase of cell cycle.

STRUCTURE OF RNA

Primary Structure of RNA

Primary structure of RNA refers to a symmetrical **phosphodiester bond backbone** similar to DNA.

Secondary Structure of RNA

Secondary structure of RNA is called **stem loop** structure or **hairpin loop** structure. When two different regions of a single RNA strand are complementary to each other they attach with one another forming a loop structure (Figure 2.5a).

Tertiary Structure of RNA

Tertiary structure of RNA is called transfer RNA or **tRNA** or **cloverleaf** structure. This is a very significant structure because it is involved in protein synthesis (translation).

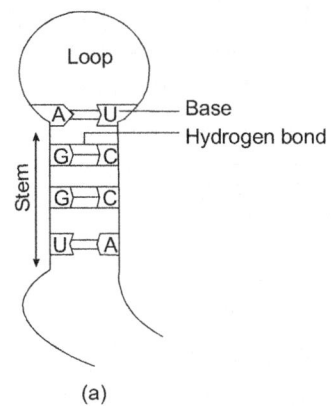

(a)

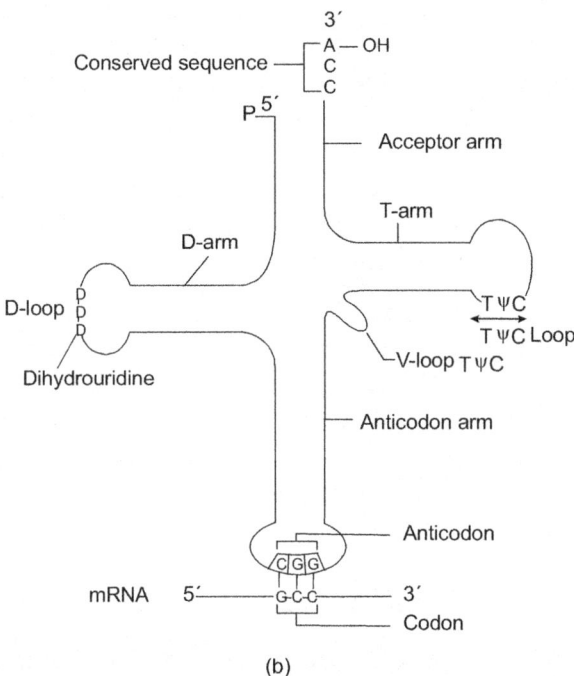

(b)

Figure 2.5   Structure of RNA (a) Secondary structure—stem loop structure (b) Tertiary structure—cloverleaf structure

RNAs can fold themselves due to complementary base-pairing forming a three looped structure called cloverleaf structure (Figure 2.5b). tRNA has four stems/arms and four loops:

- Acceptor arm
- Dihydrouridine arm (D-arm) with D-loop
- Anticodon arm
- Variable loop (V-loop)
- T ψ C arm with T ψ C loop

The acceptor arm forms the free 5′ and 3′ end of the RNA molecule. The 3′ end has a conserved (invariable) CCA sequence which is **unpaired**. This sequence is used as an identity by aminoacyl-tRNA synthetase for amino acylation during translation. This sequence is used as a primer by tRNA nucleotidyl transferase to begin synthesis of the CCA triplet. The D-arm is characterized by a **modified nucleoside** called **dihydrouridine**. This base is present in the loop of the arm, hence the name **D-loop**.

The anticodon arm has a set of three bases called anticodon which base-pairs with codon of mRNA during protein synthesis. The variable loop is a small stretch of RNA with 3–20 nucleotides lying between the anticodon arm and acceptor arm. T ψC arm has a specific base called thymidine-pseudouridine-cytosine (T ψC) present in its loop hence the name T ψC loop.

REPLICATION OF DNA

Replication is a process by which each strand of double-helical DNA is duplicated into a daughter strand in the presence of various enzymes. Thus two new strands are formed after replication. The parent strand is called **template**. A simplified representation of replication is given in Figure 2.6.

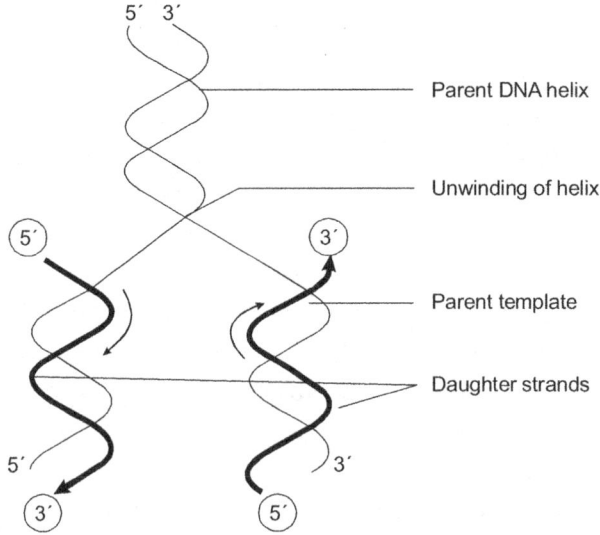

5′ 3′

Parent DNA helix

Unwinding of helix

Parent template

Daughter strands

(Arrow indicates direction of replication—5′ →3′)

Figure 2.6   Replication process

The mechanism of replication is best understood in prokaryotes. The location/sequence of DNA where the replication starts is called "origin of replication" commonly called *ori C*.

Requirements

Basic requirements for replication are as follows:

1. *Substrate* The four deoxynucleoside triphosphates (dNTPs) — dATP, dGTP, dCTP and dTTP.

2. *Enzymes* Helicase, primase, gyrase, DNA polymerase and ligase.

3. *Template strands* The two parent strands function as templates.

4. *Primer* A short strand of RNA that provides free 3′-OH end to prime the replication in 5′ to 3′ direction.

Mechanism of Replication

The mechanism of replication is illustrated in Figure 2.7 and the steps involved in the replication mechanism are summarized in Figure 2.8. Replication in *oriC* occurs by the following steps in the presence of specific enzymes.

- Template strands unwind to form a forklike structure called **replication fork**. Helicase is the enzyme responsible for catalysing the process of unwinding.

- The single-stranded templates are stabilized by a protein called **single strand binding protein** (SSB).

- **Gyrase** relaxes the positive supercoils in the helical structure.

- One of the template strands which is in 5' to 3' direction serves as template for leading strand synthesis. Leading strand is the daughter DNA produced as a continuous single strand.

- The other template which is in the 3' to 5' direction serves as template for **lagging strand** synthesis. Lagging strand is the DNA produced discontinuously in the form of many short bits or fragments called **Okazaki fragments** (named after the scientist who discovered it).

- **Primase** brings the primer to the site of synthesis.

- Complementary substrates are added to the end of the primer and polymerization takes place by the enzyme **DNA polymerase** (DNA Pol) via phosphodiester bond formation.

- Polymerization (replication) continues until termination sequences of template are encountered. At this sequence, a specific protein called **ter-binding protein** binds and prevents helicase from further unwinding DNA. This results in termination of replication.

- Thus a leading strand and a lagging strand (Okazaki fragments) are formed by replication.

- Gaps found in lagging strands (between Okazaki fragments) are sealed by an enzyme called **ligase**.

- All the enzymes and factors involved in the replication process which are described above exist in the form of a large macromolecular complex called **replisome**.

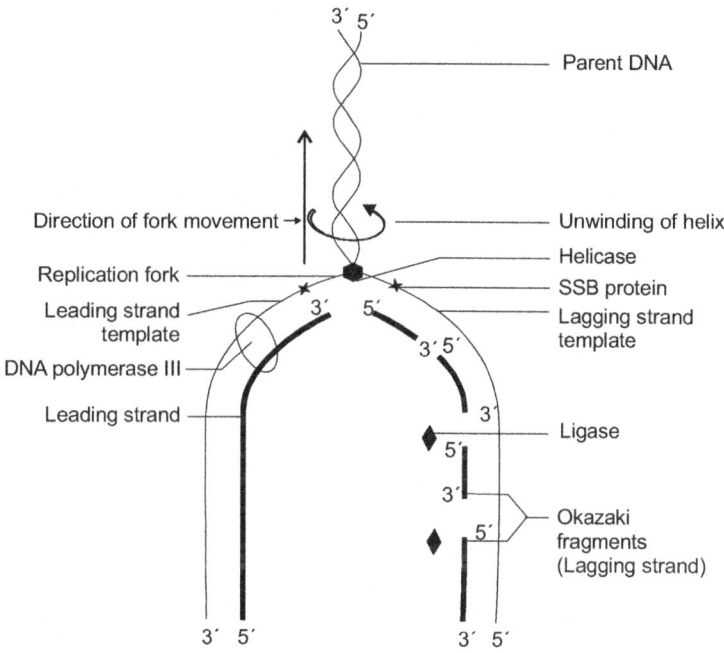

(Newly synthesized DNA is shown in bold)

Figure 2.7   Mechanism of replication

The mechanism of replication in eukaryotes is fundamentally similar to that of eukaryotes. However eukaryotic replication takes place in the presence of four types of DNA polymerases denoted as DNA Pol α, β, δ and γ. DNA Pol α is essential for the polymerization process.

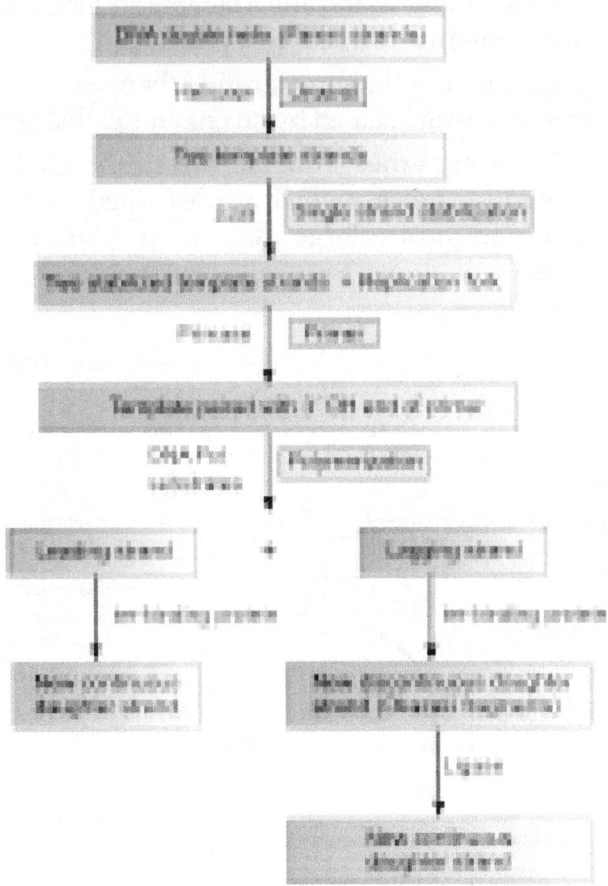

Figure 2.8 Schematic representation of steps involved in replication process

DNA Polβ plays no role in replication and only acts in DNA repair. DNA pol δ is essential for replication and it is analogous to DNA pol of leading strand synthesis of prokaryotes. DNA pol γ is needed for replication of mitochondrial DNA. Eukaryotes have two ligases, one involved in replication and the other involved in DNA repair.

Table 2.1 summarizes the functions of various enzymes and proteins involved in replication.

Table 2.1 Enzymes involved in replication

Name of the enzyme	Function
Gyrase	Relaxes the positive supercoils in the helical structure.
Helicase	Unwinds parent DNA helix to form replication fork.
SSB	Stabilizes the unwound DNA strand.
Primase	Brings the primer to the site of replication to prime DNA synthesis.
DNA Pol	Polymerizes the substrates against the template strand via phosphodiester bond formation to form new strand.
ter-binding protein	Binds and prevents helicase from further unwinding DNA resulting in termination of replication.
Ligase	Seals the gaps found in lagging strands (between Okazaki fragments)

TRANSCRIPTION

Transcription refers to the enzyme-catalysed synthesis of RNA from DNA template in $5' \rightarrow 3'$ direction. The new RNA formed from DNA is called **nascent RNA** or **primary transcript.**

Requirements

1. *Substrates*   The four nucleoside triphosphates (NTPs) – ATP, GTP, CTP, UTP.

2. *Template*   Single DNA strand. One of the strands of DNA is used as a template and is called **coding strand** while the opposite strand is called **non-coding strand** or non-template strand. Thus the new RNA formed in the transcription will have base pairs complementary to the coding strand and similar to the non-coding strand.

3. *Enzyme*   RNA polymerase (RNA pol; also called DNA-dependent RNA pol).

4. *Promoter sequence*   Specific conserved sequence that promotes accurate initiation of transcription.

    There are three types of promoter sequences:

    i.   TATAAT box (Pribnow box)

    ii. TATA box

    iii. CAAT box

5. *Transcription factors*   These are low molecular weight proteins involved in transcription process. These are of two types:

    i. *Initiation factor* σ **factor** is an initiation factor which directs RNA pol to recognize promoter sequence and initiate transcription. TFIID, TFIIE, TFIIB and TFIIA are initiation factors in eukaryotes.

    ii. *Termination factor*   **rho factor** is a termination factor which helps terminating the transcription process.

Mechanism of Transcription

The mechanism of transcription is illustrated in Figure 2.9.

The DNA helix unwinds near the promoter sequence and the two strands are separated.

RNA pol binds to the DNA, wrapping both the strands.

Substrates complementary to the template bases are added and polymerized in 5′ to 3′ direction to form new RNA.

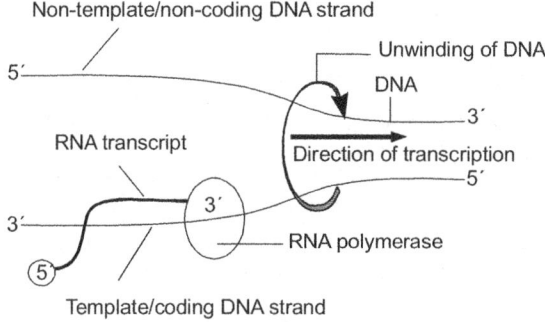

Figure 2.9 Mechanism of transcription

New RNA is called nascent RNA or primary transcript. The new RNA formed in the transcription will have base pairs complementary to the coding strand and similar to the non-coding strand.

Newly synthesized RNA extends towards the unwinding point.

The complex formed by DNA, RNA pol and new RNA is called **transcription bubble**.

As transcription continues, the transcription bubble moves along the DNA.

Transcription is terminated when the termination sequence is encountered by the bubble.

RNAs produced by transcription fall under three categories:

- mRNA/messenger RNA
- rRNA /ribosomal RNA
- tRNA/transfer RNA

These RNAs have different functions (Table 2.2). mRNA is the carrier of genetic information to be translated into proteins. rRNA forms a part of ribosomes to construct protein synthesizing machinery. tRNA carry amino acids to ribosomes for protein synthesis.

Table 2.2 Functions of different types of RNA

RNA type	Function
mRNA	Carrier of genetic information to be translated into proteins; function as template for translation
rRNA	Forms a part of ribosomes to construct protein-synthesizing machinery
tRNA	Carries amino acids to protein-synthesizing machinery

Post-Transcriptional Modification of RNA

Post-transcriptional modification refers to several structural changes that RNA undergoes to become functional. It is also called post-transcriptional processing.

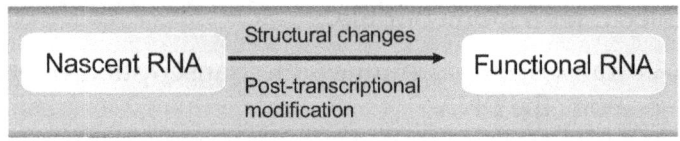

Methods for post-transcriptional modification differ between prokaryotic RNAs and eukaryotic RNAs. Removal of spacers, trimming of tRNA and addition of CCA sequence are common types of post-transcriptional modification in prokaryotes. 5' capping, 3' polyadenylation and splicing are the three types of post-transcriptional modification in eukaryotes. The different types of post-transcriptional modification are as follows.

TYPES OF POST-TRANSCRIPTIONAL MODIFICATION
Removal of spacers
Trimming of tRNA
Addition of CCA
5'capping
3'polyadenylation
Splicing

Removal of spacers Spacer sequence refers to non-functional sequences. rRNA is processed by the removal of spacer sequences. Ribonuclease P (RNase P) and ribonuclease III (RNase III) are involved in post-transcriptional modification of RNA in prokaryotes in order to render them functional. These two enzymes remove the intervening spacer sequence from the nascent prokaryotic ribosomal RNA, i.e., 30S precursor rRNA, to make it functional. This process of spacer removal is shown in Figure 2.10. 30S precursor rRNA is composed of the following components in order, in the 5' to 3' direction.

- 16S rRNA
- Spacer sequence
- tRNA
- Spacer sequence
- 23S rRNA
- Spacer sequence
- 5S rRNA

Spacer sequences flanking both ends of tRNA and those lying between 23S and 5S rRNA are cleaved by RNase P and RNase III. This process of spacer removal results in the release of 16S RNA, tRNA, 23S RNA and 5S RNA. Similarly 45S precursor rRNA of eukaryotes is composed of the following sequences in order, in the 5′ to 3′ direction:

18S RNA
Spacer sequence
5.8S RNA
Spacer sequence
28S RNA

Spacers are removed by endonucleolytic cleavage by endonucleases.

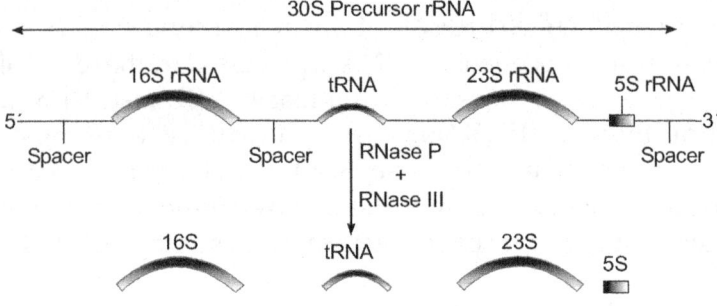

Figure 2.10   Removal of spacers in RNA

Trimming of tRNA   Newly synthesized tRNA has a non-functional sequence at its tip which should be trimmed. RNase D and RNase P are involved in trimming the terminal sequence (Figure 2.11) in order to form active functional tRNA.

Addition of CCA   Some of the tRNAs do not possess the CCA sequence at their 3′ end which is common to all tRNAs. Additon of CCA sequence to 3′ end is another mode of post-transcriptional modification. **tRNA**

nucleotidyl transferase catalyses the addition of CCA making the tRNA functional (Figure 2.12).

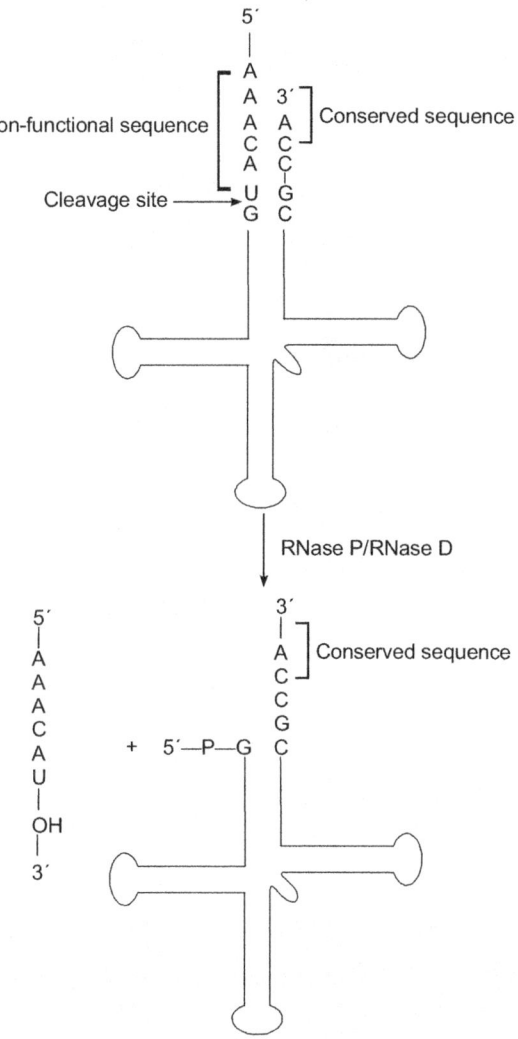

Figure 2.11   Enzymatic trimming of tRNA

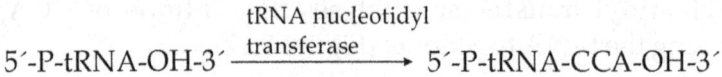

$$5'\text{-P-tRNA-OH-}3' \xrightarrow{\text{tRNA nucleotidyl transferase}} 5'\text{-P-tRNA-CCA-OH-}3'$$

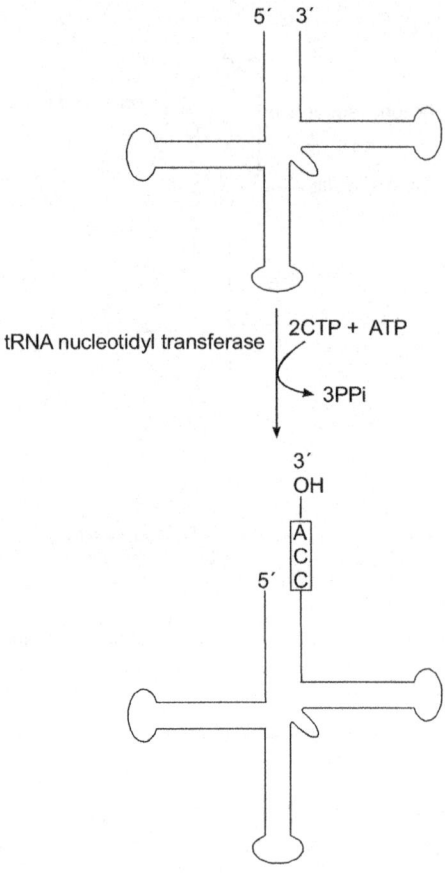

Figure 2.12   Addition of CCA to tRNA

5′ Capping   5′ capping is the common method of post-transcriptional modification in eukaryotes. Introduction of a **7-methylguanylate** (7-MG) to the 5′ end of mRNAs via 5′ to 5′ triphosphate linkage is called 5′ capping.

5´-P----------------- OH-3´———→ 5-7MG ----------------- OH-3´
Native mRNA   Stable mRNA
                          with 7MG cap

Functions of 5' *capping*   5' capping **stabilizes** mRNAs and results in more efficient translation.

3' *-Polyadenylation*   A polymer of 200–300 adenylate residues (-AAAAAA-) linked via phosphodiester bond is called poly(A) tail. Addition of poly(A) tail to 3' end of mRNA of eukaryotes by the enzyme **polyadenylyl transferase** is called 3'-polyadenylation. This process results in **stabilization** of mRNA.

5´-P----------OH-3´ ———→ 5´-P----------OH-3´-AAAAAA—
Native mRNA   Stable mRNA with poly(A) tail

Splicing   mRNA has two types of sequences: coding sequences (also called exons) which are expressed or translated into proteins and non-coding sequences (also called introns or intervening sequence) which are not expressed or converted to proteins. The process of removing the **noncoding sequences** after transcription and before translation is called splicing. Introns are characterized by the presence of guanine-uracil (GU) sequence on the 5' border and adenine-guanine (AG) sequence on the 3' border of the intron–exon junction. Splicing is catalysed by a large ribonucleoprotein complex called **spliceosome**. Spliceosome recognizes the introns by GU and AG borders and removes them resulting in the formation of a completely expressible mRNA for consequent translation.

TRANSLATION

Translation is a process by which the information encoded in mRNA is converted into proteins by enzymes, ribosomes and tRNA. It takes place in three steps, **initiation, elongation** and **termination**.

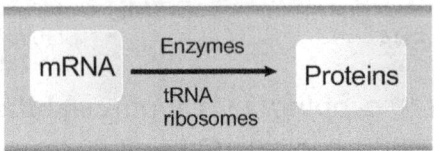

Codons

Translation occurs by assembling amino acids in a protein-synthesizing machinery constituted by ribosomes and tRNA. A set of three mRNA nucleotides are collectively called as a codon. Each codon is specific for a particular amino acid. For example, codon UCU corresponds to corresponds to serine, codon CUU corresponds to proline and codon GUU corresponds to valine.

Components Needed for Translation

Various components needed for translation process are:

- 30S ribosomal subunit containing 16S rRNA (for prokaryotes), 40S ribosomal subunit containing 18S rRNA (for eukaryotes)
- mRNA to be translated
- Formylmethionyl-tRNA (Fmet-tRNA)
- Aminoacyl-tRNA synthetase
- ATP
- Transformylase
- GTP
- Peptidyl transferase
- Release factors

Mechanism of Translation

Mechanism of translation involves three stages:

- Initiation
- Elongation
- Termination

Translation process is schematically described in Figure 2.13 and diagrammatically represented in Figure 2.14.

30S initiation complex is needed for initiation of translation. This is formed by binding of pyrimidine sequence of 16S rRNA component of **30S subunit** with the purine-rich sequence of mRNA which is called **Shine–Dalgarno sequence.**

tRNA is converted to methionyl-tRNA (met-tRNA) by aminoacyl-tRNA synthetase at the expense of ATP.

met-tRNA is converted to formylmethionyl-tRNA (Fmet-tRNA or initiator tRNA) by transformylase.

Finally, initiator tRNA binds to the 30S complex.

The 30S complex then binds to **50S subunit** of the ribosome forming **70S complex.**

70S subunit has two sites, **A site** or the amino site and **P site** or the peptide site. Initiator tRNA carries methionine and has anticodon (UAC) which is complementary to the start codon AUG of mRNA at the P site. Thus, initiator tRNA carrying methionine is attached to the P site.

ELONGATION

Let us consider for convenience that the codons lying to the right of the start codon (AUG) are as follows:

CAU (specific for histidine)

GGG (specific for glycine)

CAG (specific for glutamine)

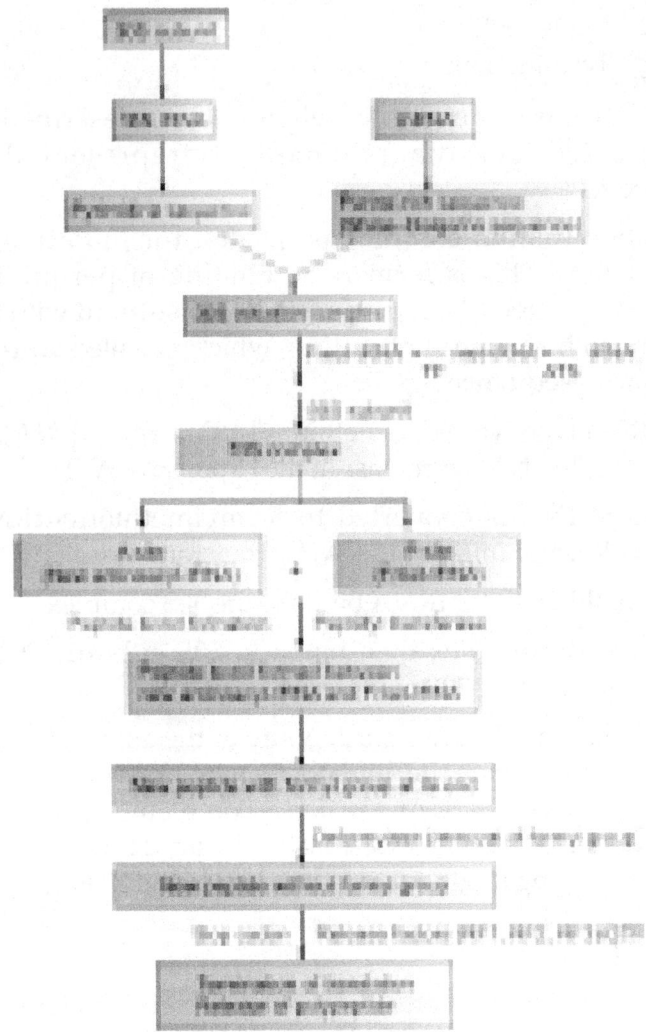

Figure 2.13 Schematic representation of translation process (TF—transformylase; ATS—aminoacyl-tRNA synthetase)

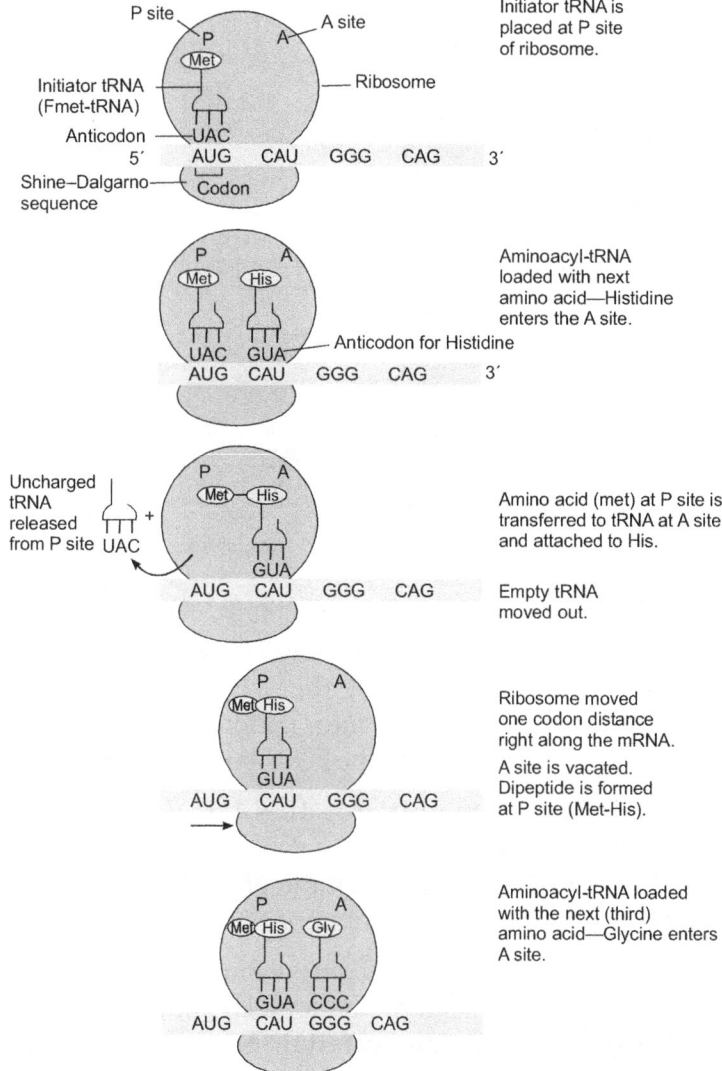

Figure 2.14 Diagrammatic representation of translation process (*Continues*)

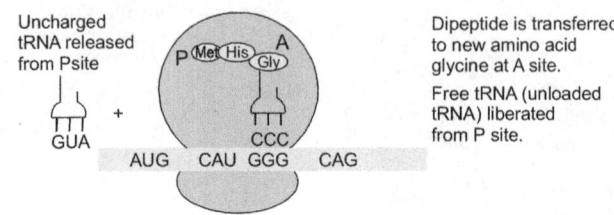

Uncharged tRNA released from Psite

GUA

Dipeptide is transferred to new amino acid glycine at A site.

Free tRNA (unloaded tRNA) liberated from P site.

Entry of new amino acid to A site and transfer of peptide to new amino acid continues as a cycle.

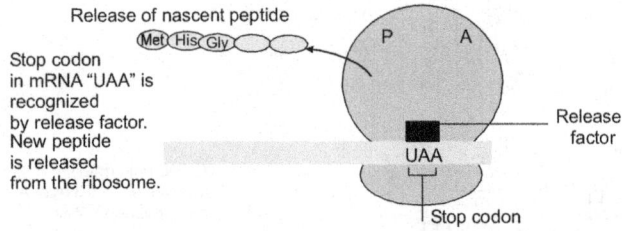

Release of nascent peptide

Stop codon in mRNA "UAA" is recognized by release factor. New peptide is released from the ribosome.

Release factor

Stop codon

Figure 2.14   Diagrammatic representation of translation process

The first incoming aminoacylated tRNA (in the above example, it is tRNA with His), that is complementary to the codon adjacent to the initiator codon (CAU), binds to the A site (the amino site) of the 70S ribosomal subunit.

Methionine from initiator tRNA in the P site is transferred to histidine at A site. Peptide bond is formed between the two amino acids in the presence of **peptidyl transferase**, thus starting elongation.

This enzyme has an RNA component which is actually responsible for the catalytic action. Therefore, peptidyl transferase is a **ribozyme (catalytic RNA or enzymatic RNA)**.

Simultaneously, the uncharged initiator tRNA relieves out of the ribosome vacating the P site.

Then the ribosome moves along three nucleotides in 5′ – 3′ direction along the mRNA (i.e., moves one codon distance to the right).

As a result, methionine–histidine dipeptide will be at the P site, and A site will be vacant.

Now, the next amino acid glycine comes with its tRNA to the A site.

Then the dipeptide is transferred to glycine and attached via peptide bond as before. Uncharged rRNA is released.

Ribosome moves one codon distance further leaving a free A site.

tRNA with glutamine enters A site. Tripeptide at P site moves to glutamine and peptide bond is formed.

Elongation of peptide chain continues.

TERMINATION

Elongation continues until the termination sequence or stop codon denoted as UAA is encountered. Release factors like RF1, RF2, RF3 are involved in termination process. RF1 and RF2 recognize stop codons like UAA, UAG and UGA.

RF3 directs peptidyl transferase to release the polypeptide from the tRNA in the P site by hydrolysis in the presence of GTP.

Deformylase **removes** the **formyl** group from the growing polypeptide chain.

Then the ribosomal subunit is also dissociated at the expense of GTP hydrolysis.

Functions of enzymes catalysing translation are summarized in Table 2.3.

Table 2.3 Enzymes catalysing translation

Enzyme	Function
Aminoacyl - tRNA synthetase	Catalyses the esterification of a specific amino acid to form an aminoacyl-tRNA at the expense of ATP
Transformylase	Catalyses the transfer of formyl group to the methionine which is carried by the tRNA
Peptidyl transferase	Catalyses the formation of peptide bond between adjacent amino acids in the ribosome machinery
Deformylase	Removes the formyl group from the growing polypeptide chain.

Post-translational Modification

Post-translational modification refers to some structural changes that take place in native protein in order to render them functional and biologically active.

Three types of post-translational modification are common in proteins:

i. Formation of disulphide bridges

ii. Glycosylation

iii. Proteolytic cleavage

i. *Formation of disulphide bridge* **Protein disulphide isomerase** catalyses the formation of disulphide bridges at certain parts in the peptide chain where there are cysteine residues. This results in the conversion of protein or peptide chain into properly folded form which forms the tertiary and quaternary conformation (Figure 2.15).

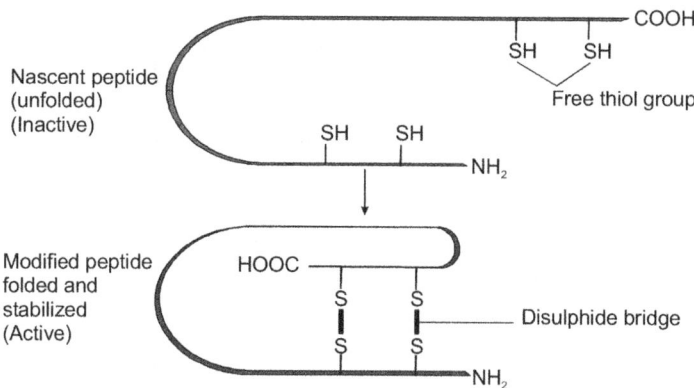

Figure 2.15 Post-translational modification by formation of disulphide bridge

ii. *Glycosylation* Glycosylation means addition of carbohydrates in the form of oligosaccharides to the proteins. This process is catalysed by an enzyme called **glycosyl transferase.** This enzyme recognizes asparagine residues lying in the vicinity of serine or threonine in the protein and attaches the oligosaccharides (Figure 2.16).

iii. *Proteolytic cleavage* Proteolytic cleavage is the well known method of post-translational modification in hormones and enzymes. Some enzymes are synthesized as biologically inactive proteins called zymogens or proenzymes. Example, chymotrypsin (protein-digesting enzyme) is derived from chymotrypsinogen (inactive form of chymotrypsin) and carboxypeptidase is derived from its inactive precursor procarboxypeptidase. Proelastase is the zymogen for elastase. Pepsinogen is the inactive precursor for pepsin. Similarly inactive hormones are called pre-prohormones or prohormones. Insulin the active hormone is derived from proteolytic cleavage of pre-proinsulin and pro-insulin in pancreatic cells.

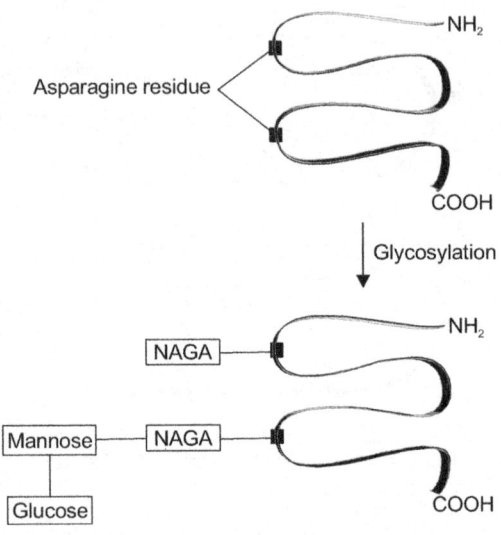

NAGA—*N*-Acetylglucosamine

Figure 2.16   Post-translational modification by glycosylation

Example 1

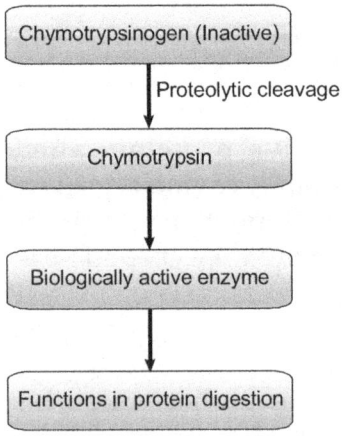

Example 2

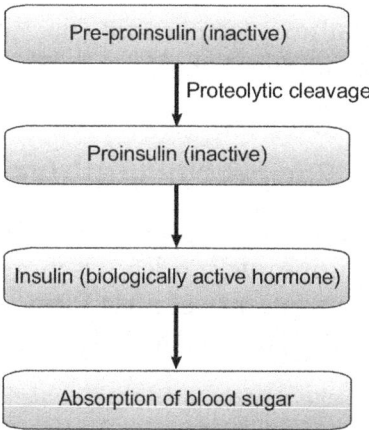

Example 3

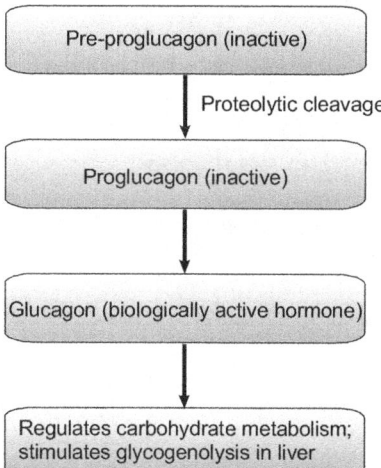

METABOLISM OF NUCLEOTIDES

Purine nucleotides (AMP and GMP) and pyrimidine nucleotides (TMP, CMP, UMP) are synthesized through two pathways:

1. De novo pathway (synthesizing nucleotides from the basic raw materials).
2. Salvage pathway (synthesizing nucleotides from the metabolic intermediates).

De novo Pathway for the Synthesis of Purine Nucleotides

PRPP — Phosphoribosyl pyrophosphate is the precursor for the synthesis of purine nucleotides. It is formed from ribose 5-phosphate in the presence of ATP. PRPP is converted to IMP which in turn is converted to AMP and GMP as shown below:

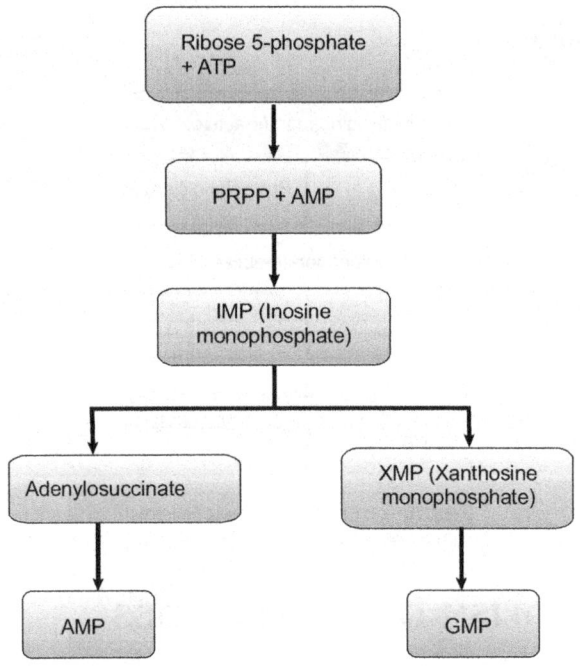

The steps involved in de novo pathway are shown in Figure 2.17.

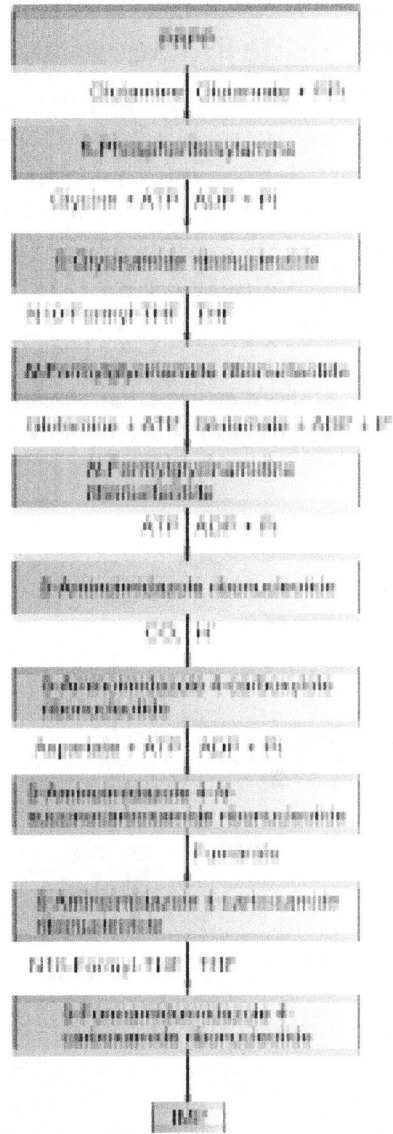

Figure 2.17 De novo pathway for the synthesis of purine nucleotides (AMP and GMP)

Salvage Pathway for the Synthesis of Purine Nucleotides

GMP is formed from guanine base and PRPP in the presence of hypoxanthine-guanine phosphoribosyltransferase **(HGPRT)**. HGPRT catalyses the formation of GMP from hypoxanthine too. AMP is formed from adenine base and PRPP in the presence of adenine phosphoribosyltransferase (APRT). The salvage pathway for the synthesis of purine nucleotides is shown in Figure 2.18.

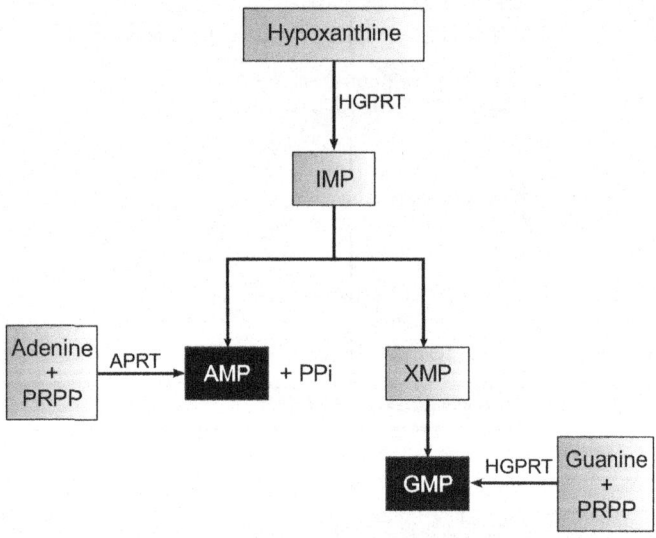

Figure 2.18   Salvage pathway for the synthesis of purine nucleotides

De novo Pathway for the Synthesis of Pyrimidine Nucleotides

In de novo pathway pyrimidine nucleotides are synthesized starting from glutamine and carbon dioxide as shown in Figure 2.19.

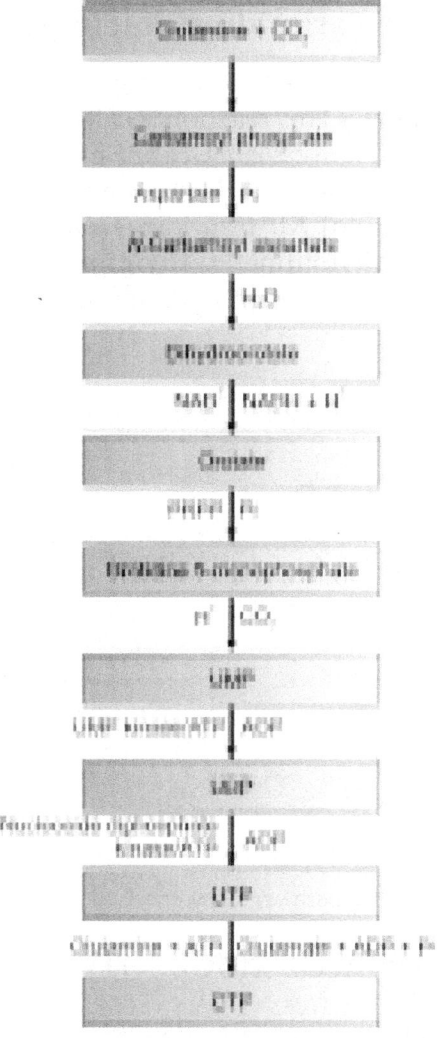

Figure 2.19   De novo pathway for synthesis of pyrimidine nucleotides

Salvage Pathway for the Synthesis of Pyrimidine Nucleotides

Uracil base is converted to uridine in the presence of ribose 1-phosphate. This reaction is catalysed by **uridine-phosphorylase** (Figure 2.20). Uridine in turn is converted to UMP by uridine kinase in the presence of ATP. Cytidine can be converted into CMP by uridine kinase. Thymine base is converted to thymidine by **thymine phosphorylase; thymidine kinase** converts thymidine to dTMP.

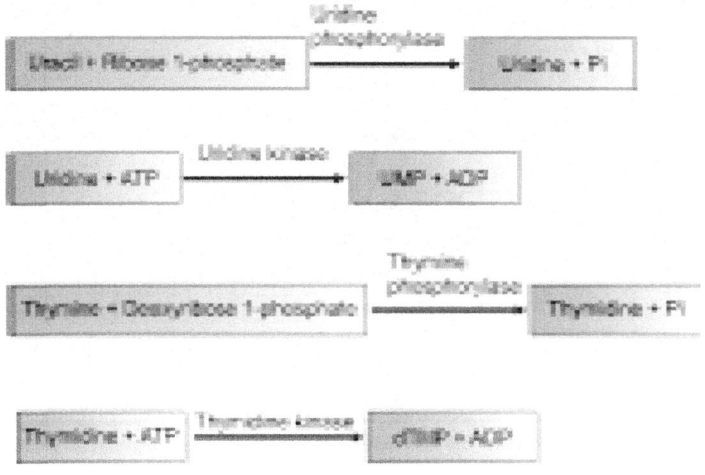

Figure 2.20   Salvage pathway for synthesis of pyrimidine nucleotides

DEGRADATION OF NUCLEOTIDES

Purine and pyrimidine nucleotides are first converted into their nucleosides and then into free bases. Purine bases are converted to uric acid. Pyrimidine bases are converted into α-alanine and β-aminoisobutyrate (Figure 2.21).

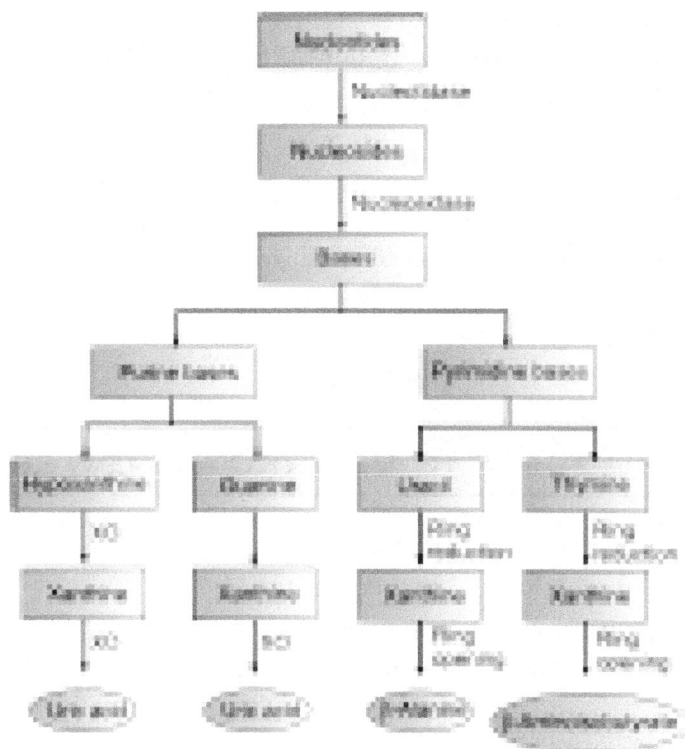

Figure 2.21   Degradation of nucleotides (XO—xanthine oxidase)

REVIEW QUESTIONS

1. What are nucleotides?
2. What is phosphodiester bond?
3. What is Chargaff's rule?
4. What is replication?
5. What are Okazaki fragments?
6. Mention the functions of various enzymes involved in replication.

7. What is transcription?
8. Write the function of mRNA, tRNA and rRNA.
9. Define poly(A) tail and polyadenylation.
10. What is translation?
11. Write the functions of various enzymes involved in translation process.
12. What do you mean by de novo synthesis? Give an example.
13. What are the metabolic end products of purine and pyrimidine nucleotides?

CHOOSE THE CORRECT ANSWER

1. The combination of sugar and base is called
   (a) nucleoside   (b) nucleotide
   (c) nucleic acid   (d) nucleolus
2. Double helix structure of DNA is also called
   (a) Watson and Crick model
   (b) secondary structure of DNA
   (c) tertiary structure of DNA
   (d) cloverleaf model
3. Which of the following enzyme joins Okazaki fragments?
   (a) helicase   (b) primase
   (c) ligase   (d) gyrase
4. D-loop is present in
   (a) mRNA   (b) tRNA
   (c) rRNA   (c) DNA
5. Which of the following is not a promoter sequence?
   (a) TATAAT box (Pribnow box) (b) TATA box
   (c) CAAT box   (d) CCATT box

In greek, *proto* means "foremost" and hence, proteins are the foremost macromolecules in the biological system. Proteins are linear unbranched polymers of twenty different α-**amino acids** that are encoded in the DNA of the genome. Proteins are synthesized in the cytoplasm from RNA by a process called **translation**.

AMINO ACIDS—BUILDING BLOCKS OF PROTEINS

Amino acids form the building blocks of proteins. The general structure of an amino acid is represented in Figure 3.1. It has a central carbon which is called α-carbon. The α-carbon is linked to four groups:

1. Amino group ($-NH_2$)
2. Carboxyl group ($-COOH$)

3. Hydrogen atom($-$H)
4. Variable group ($-$R) which is specific for each amino acid.

If the amino group is on the right side of the α-carbon, it is called D-amino acid. If the amino group is on the left side of the α-carbon, it is called L-amino acid. For example, L-alanine and D-alanine are shown in Figure 3.2. L-amino acids are biologically active. L-amino acids are incorporated into proteins while D-amino acids are found in many bacterial products and peptide antibiotics. Except for glycine, the central carbon of all other amino acids are linked to four different groups and that is called chiral carbon. Chiral carbon gives rise to **optical activity.**

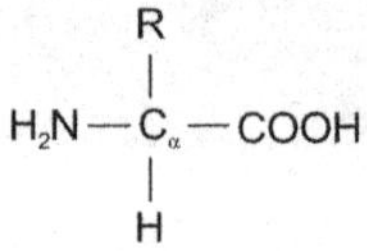

$$H_2N - C_\alpha - COOH$$

with R above and H below the C_α

Figure 3.1   General structure of amino acid

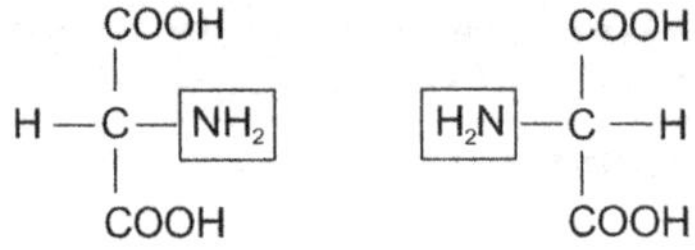

COOH
|
H — C — NH₂
|
COOH

(a)

COOH
|
H₂N — C — H
|
COOH

(b)

Figure 3.2   Structure of (a) D-alanine and (b) L-alanine

There are twenty possible variable groups giving rise to twenty amino acids. Twenty amino acids can conventionally be grouped under different families. The family of the amino acid, name of the amino acid and the structure of twenty amino acids are given in Figure 3.3.

(a) Amino acids with hydrophobic side groups

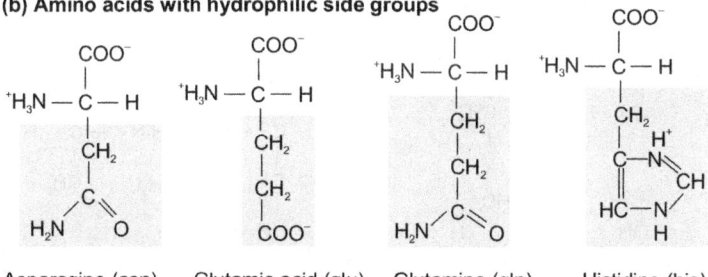

Figure 3.3   Family, name and structure of the twenty amino acids (*Continues*)

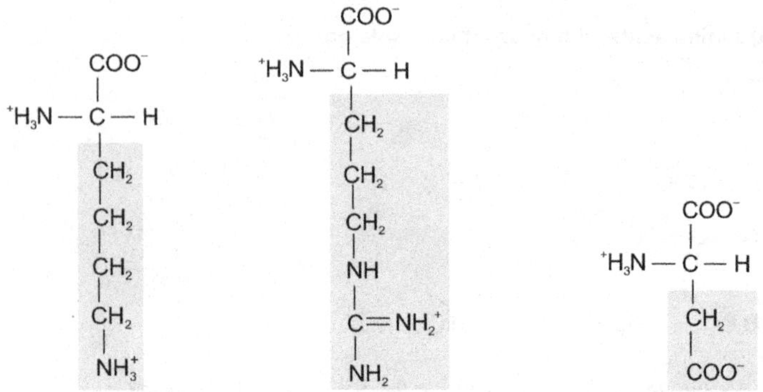

Lysine (lys)   Arginine (arg)   Aspartic acid (asp)

(c) Amino acids that are in between

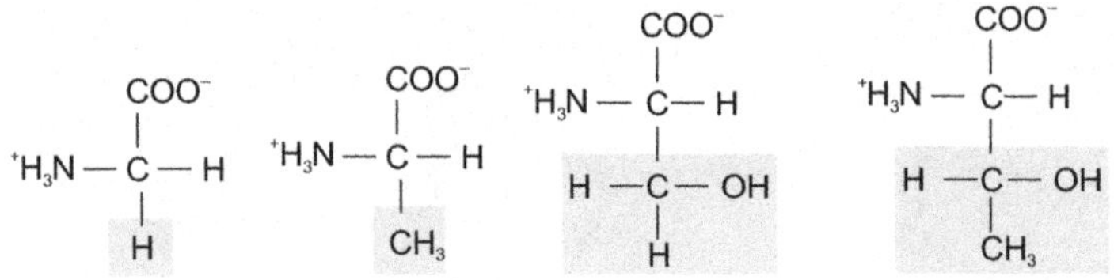

Glycine (gly)   Alanine (ala)   Serine (ser)   Threonine (thr)

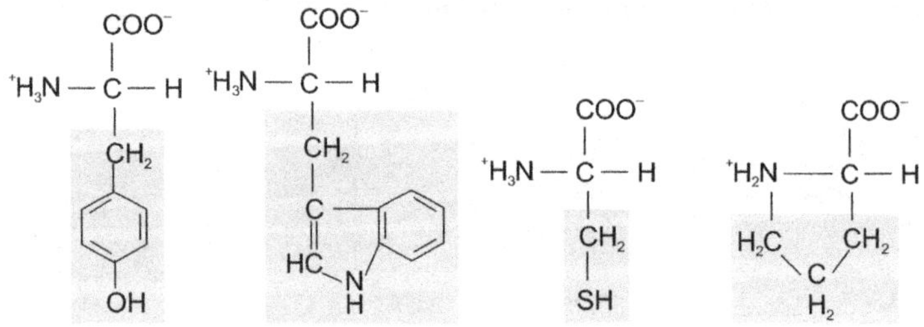

Tyrosine (tyr)   Tryptophan (trp)   Cysteine (cys)   Proline (pro)

Figure 3.3   Structures of twenty amino acids

CHARGE DISTRIBUTION IN AMINO ACIDS

In aqueous solution, the carboxyl group of the amino acid is dissociated and negatively charged while the amino group is protonated and positively charged. Thus, amino acids exist as dipolar molecules. Such molecules which are electrically neutral with no net charge are called zwitterions. But in acidic and alkaline solutions, amino acids bear a net positive or negative charge. In acidic solution, i.e., at low pH, the carboxyl group of the amino acid accepts a proton and becomes uncharged making the molecule positive. In alkaline solution, i.e., at high pH, the amino group of the amino acid becomes uncharged making the molecule negative. Zwitterionic form and charged forms of amino acids are shown in Figure 3.4.

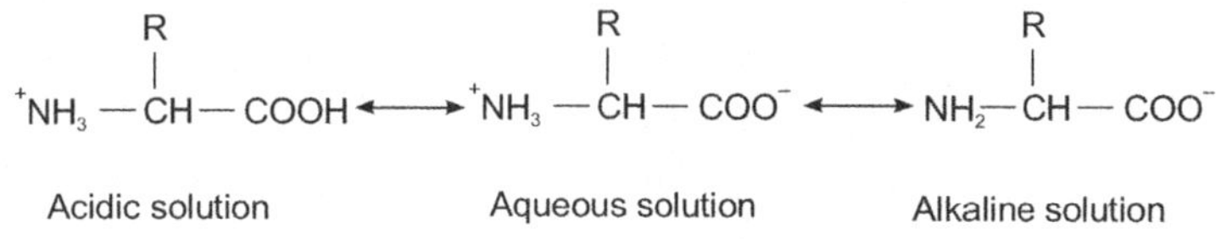

$$\overset{\displaystyle R}{\underset{\displaystyle |}{}} \quad \overset{\displaystyle R}{\underset{\displaystyle |}{}} \quad \overset{\displaystyle R}{\underset{\displaystyle |}{}}$$

$$^+NH_3-CH-COOH \longleftrightarrow {}^+NH_3-CH-COO^- \longleftrightarrow NH_2-CH-COO^-$$

Acidic solution	Aqueous solution	Alkaline solution

Figure 3.4 Charge distribution in amino acids in different solutions

Substances which have a dual nature, i.e., which can act both as an acid (proton donor) or as a base (proton acceptor) are called amphoteric substances or **ampholytes**. Amino acids are amphoteric in nature because the zwitterionic form of amino acid can donate proton or accept proton as shown below:

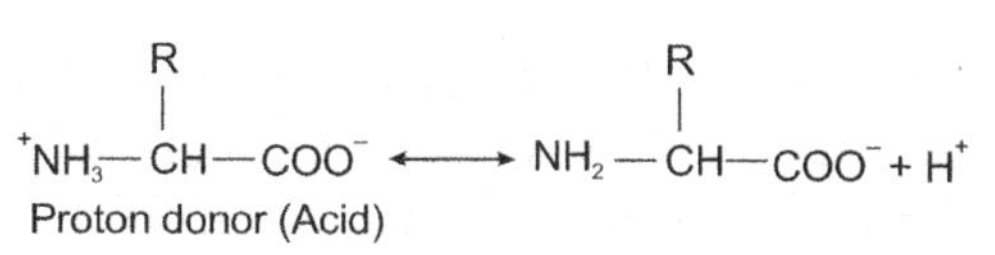

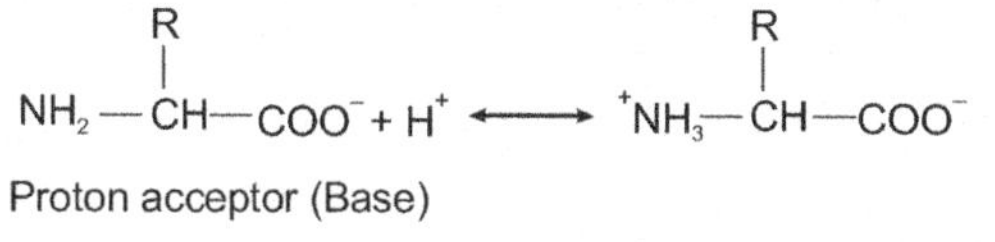

$$^+NH_3-CH-COO^- \longleftrightarrow NH_2-CH-COO^- + H^+$$

Proton donor (Acid)

$$NH_2-CH-COO^- + H^+ \longleftrightarrow {}^+NH_3-CH-COO^-$$

Proton acceptor (Base)

ESSENTIAL AND NONESSENTIAL AMINO ACIDS

Out of the twenty amino acids, the human body can synthesize only twelve amino acids. The rest eight amino acids should be provided in the diet and are called essential amino acids. Table 3.1 lists the eight essential amino acids.

Table 3.1   Essential amino acids (eight out of twenty)

Name of the essential amino acid
Threonine
Tryptophan
Methionine
Isoleucine
Leucine
Lysine
Valine
Phenylalanine

NON-STANDARD AMINO ACIDS

Apart from the 20 standard amino acids mentioned in Figure 3.3, other amino acids called non-standard amino acids are also found in some proteins. These amino acids are usually modified forms of standard amino acids. 4-hydroxyproline, 5-hydroxyproline, *N*-methyllysine, γ-carboxyglutamate, desmosine and selenocysteine are well known examples for non-standard amino acids. The precursors of these amino acids and their sites of occurrence are given in Table 3.2.

Table 3.2 Non-standard amino acids

Non-standard amino acid	Precursor	Occurrence
4-hydroxyproline	Proline	Plant cell wall protein
5-hydroxyproline	Proline	Collagen of connective tissue
N-methyllysine	Lysine	Myosin – muscle protein
γ-carboxyglutamate	Glutamate	Prothrombin – blood clotting protein
Desmosine	Lysine	Elastin – fibrous protein
Selenocysteine	Cysteine	Glutathione peroxidase – detoxification enzyme

LINKING OF AMINO ACIDS

Amino acids are linked to one another by peptide bonds. The amino group of one amino acid reacts with carboxyl group of another amino acid to form a peptide bond with the generation of water molecule. Formation of peptide bond is shown in Figure 3.5.

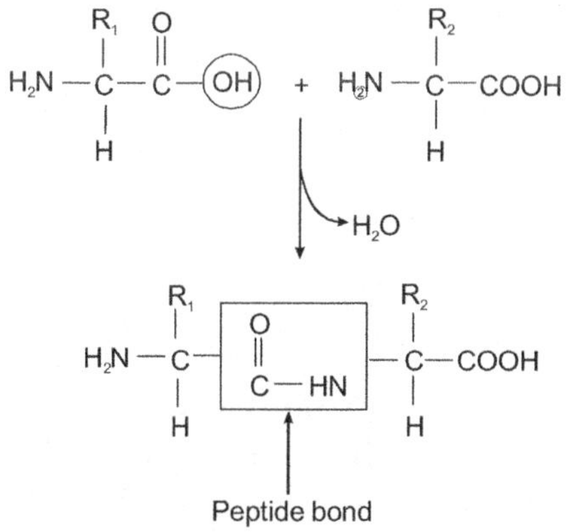

Figure 3.5 Formation of peptide bond between two amino acids

PROPERTIES OF AMINO ACIDS

Physical Properties

1. Amino acids are white and crystalline.
2. They are generally soluble in water and insoluble in organic solvents.
3. They have high melting point.
4. They are optically active, except glycine because it possesses chiral carbon.
5. They are ampholytes.

Chemical Properties

1. **Ninhydrin reaction**   All amino acids except proline react with ninhydrin (chemically called hydrindene hydrate) to form a purple pigment. Hence ninhydrin test is used to identify amino acids. Ninhydrin test is given below.

2. **Sanger's reaction**   Amino acids react with Sanger's reagent (named after the scientist who discovered the reaction; chemically called 1-fluoro 2,4-dinitrobenzene; under mild alkaline condition to give dinitrophenyl

amino acid which forms a yellow colour. This is a condensation reaction and it is illustrated below:

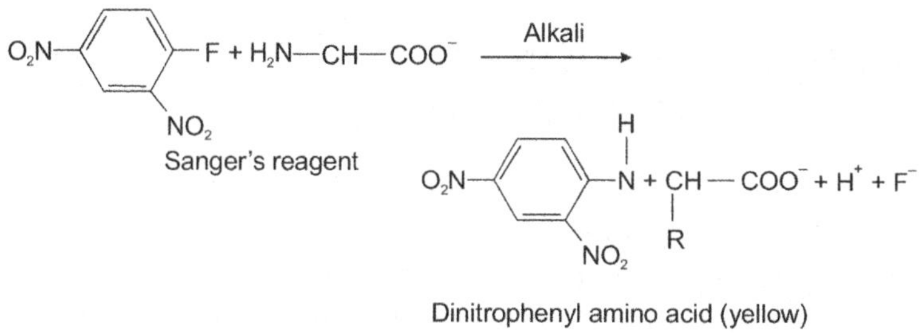

Dinitrophenyl amino acid (yellow)

3. Reaction with formaldehyde   Amino acids react with formaldehyde forming hydroxymethyl derivative.

4. Formation of Schiff's base   Amino acids form Schiff's base with benzaldehyde.

5. van Slyke reaction   Reaction of amino acids with nitrous acid to form nitrogen gas and alpha hydroxy acid is called van Slyke reaction.

6. Salt formation   Amino acids react with sodium hydroxide forming sodium salts which are of commercial importance.

7. Esterification reaction or Emil Fischer reaction   Carboxyl group of amino acids react with alcohols to form corresponding esters.

8. Reaction with amine   Carboxyl group of amino acids react with amines to produce amide.

9. Edman reaction   Phenyl isothiocyanate (Edman reagent) reacts with the amino group and carboxyl group of amino acids to form thiohydantoic acid which in the presence of nitromethane under acidic condition cyclizes to form thiohydantoin.

10. Reaction with phosgene Reaction of amino acid with phosgene produces N-carboxy anhydride and hydrochloric acid.

COLOUR REACTIONS FOR SPECIFIC AMINO ACIDS

All reactions described under chemical properties are common for all amino acids. But some reactions, usually colour reactions, are unique for specific amino acids. These colour reactions are used to identify specific amino acids and are discussed here.

1. *Xanthoproteic test for aromatic amino acids* Xanthoproteic test is specific for amino acids with benzene rings. Thus this test is characteristic of tyrosine, tryptophan and phenylalanine. These amino acids react with concentrated nitric acid to produce yellow colour on boiling. The yellow colour deepens and turns orange upon addition of an alkali like ammonia or sodium hydroxide.

2. *Millon's test for tyrosine* Millon's test involves addition of 10% mercuric sulphate prepared in 10% sulphuric acid to amino acids, heating for half a minute followed by cooling and addition of 1% sodium nitrite. This results in the formation of red colour which is specific for tyrosine.

3. *Aldehyde test for tryptophan* When treated with a solution of mercuric sulphate in sulphuric acid, the indole nucleus of tryptophan is oxidized which then reacts with formaldehyde to form a violet-coloured complex.

4. *Sakaguchi's test for arginine* The guanidine group of arginine reacts with α-naphthol under alkaline condition to form carmine a red colour.

5. *Pauly's test for tyrosine and histidine* Tyrosine and histidine react with diazotized sulphanilic acid in alkaline condition to form red colour.

6. Ehrlich test for tryptophan Tryptophan forms blue colour with p-dimethylamino benzaldehyde under acidic condition.

STRUCTURE OF PROTEINS

Proteins display four different levels of structural organization.

- Primary structure
- Secondary structure and supersecondary structure
- Tertiary structure
- Quaternary structure

Examples for these structures and factors stabilizing these structures are summarized in Table 3.3. A diagrammatic representation of the four levels of structural organization on proteins are shown in Figure 3.6.

Table 3.3 Levels of structural organization of proteins

Structure	Stabilizing factor	Example
Primary	Peptide bond	Glutathione
	Disulphide bridge	Insulin
Secondary	Hydrogen bond	Keratin
Tertiary	Hydrogen bond	Myoglobin
	Hydrophobic interaction	
	van der Waals interaction	
	Ionic bond	
Quaternary	Hydrogen bond	Haemoglobin
	Hydrophobic interaction	
	van der Waals interaction	
	Ionic bond	

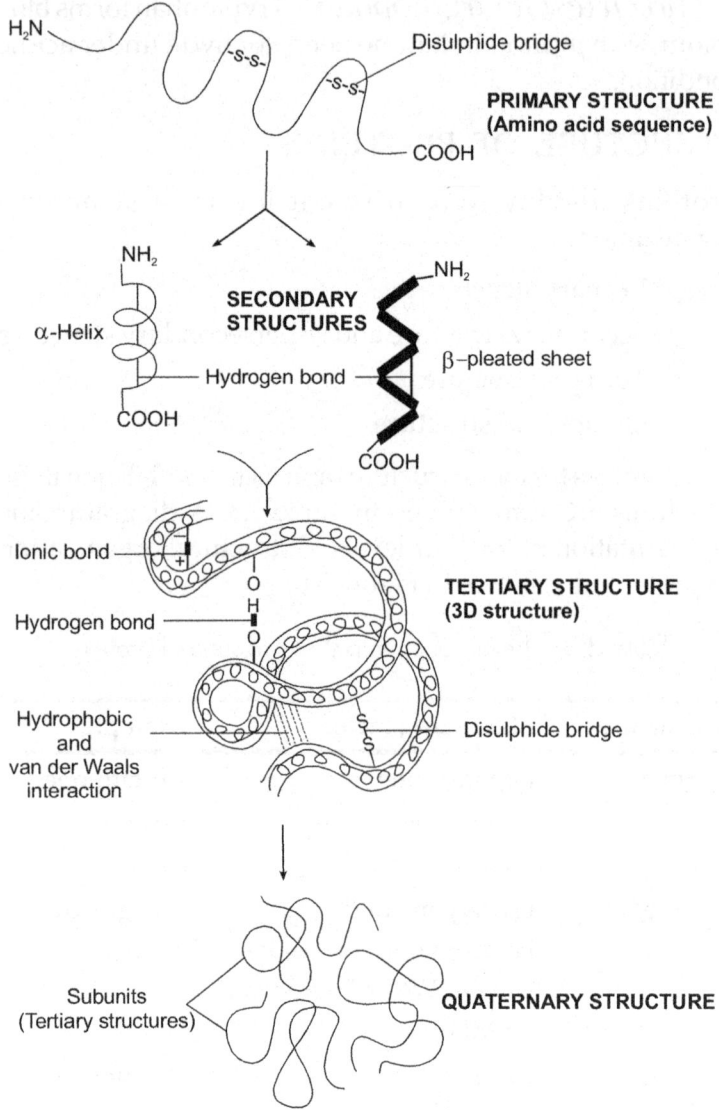

Figure 3.6   Structural organization of proteins

Primary Structure

Primary structure refers to the amino acid sequence of the peptide chain with amino group at the left end and carboxyl group at the right end. An example for protein with primary structure is the tripeptide glutathione. The sequence of amino acids in the peptide chain is always read from the N-terminus to C-terminus (i.e., from left to right). This structure is stabilized by the peptide bond and disulphide bond.

Secondary Structure

Secondary structure refers to simple foldings and bendings of peptide chains (the primary structure). Two types of secondary structures are known. **Helix,** in which the peptide chain is coiled and β-pleated sheet in which the peptide chain is bent to form a zig-zag conformation. If a pair of β-**pleated sheets** have their amino terminal end in opposite direction they are called **anti-parallel** sheets. Helix and pleated sheet structures are stabilized by **hydrogen bond** formed between nitrogen atom and oxygen atom of different amino acid residues within the chain. In the secondary structure, all the R groups will be facing outwards. Keratin is an example for secondary structure.

Supersecondary structure Secondary structure elements are observed to combine in specific geometric arrangements known as motifs or supersecondary structures. There are four types of supersecondary structures in proteins (Figure 3.7). They are:

- β-hairpin turn — two anti-parallel sheets linked by hairpin loop.
- β-meander — three antiparallel sheets linked by hairpin loop.

- α-hairpin — two helices linked by hairpin loop.
- α-β-motif — two antiparallel sheets linked by a helix.

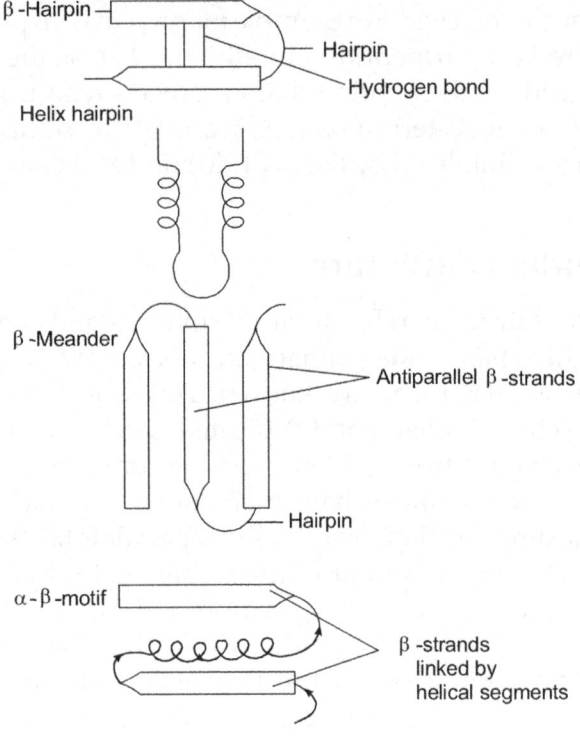

Figure 3.7 Supersecondary structures of proteins

Tertiary Structure

Tertiary structure is the **three-dimensional** structure of a completely **folded** and **compacted** single protein molecule. This may have both helical and pleated-sheet conformation. It represents the spatial arrangement of the secondary structures. Tertiary structure is stabilized by hydrogen bond, salt bridge and van der Waals force. Examples, are myoglobin and dihydrofolate reductase.

Quaternary Structure

Quaternary structure is the assembly of two or more polypeptide chains which are individually folded and compacted. Each chain in the assembly is called as a **subunit**. An example is haemoglobin made up of four subunits.

INTERACTIONS IN PROTEIN STRUCTURE

Interactions between atoms falls into two categories: covalent interaction and non-covalent interaction. Covalent interaction is more stronger and it is formed due to sharing of electrons between atoms. Non-covalent interaction is generally weak and it is due to van der Waals interaction, hydrogen bonding and electrostatic bonding. van der Waals interaction is a weak interaction created due to dipole–dipole interactions formed via transient inhomogeneities in the electrons within a molecule. Hydrogen bond is formed due to the weak sharing of an electron pair between a hydrogen atom and another atom. Electrostatic bonds are the result of the attraction between groups with opposite electrical charges.

Covalent and non-covalent interactions in protein molecules are responsible for maintaining the secondary, tertiary and quaternary structure of proteins. Interactions in protein structure can be categorized as follows:

Covalent Interaction

Covalent interaction includes peptide bond and disulphide bond.

Peptide bond Peptide bond is the covalent bond created by a dehydration reaction that occurs when two amino acids are positioned so that the carboxyl group of one is adjacent to the amino group of the other (Figure 3.8a).

$$\text{H}_2\text{N—Amino acid}_1\text{—COOH} + \text{H}_2\text{N—Amino acid}_2\text{—COOH}$$
$$\downarrow \text{H}_2\text{O}$$
$$\text{H}_2\text{N—Amino acid}_1\text{—CONH—Amino acid}_2\text{—COOH}$$

Disulphide bond   Disulphide bond is formed between two thiol groups (—SH) possessed by two different cysteine residues (Figure 3.8b).

- This is the strongest chemical bond present in protein structure. For example, insulin is stabilized by disulphide bridges.

Protein— Cysteine — SH + HS— Cysteine — Protein
(Free thiol)   (Free thiol)

Protein—Cysteine —S—S— Cysteine—Protein

Disulphide
bridge

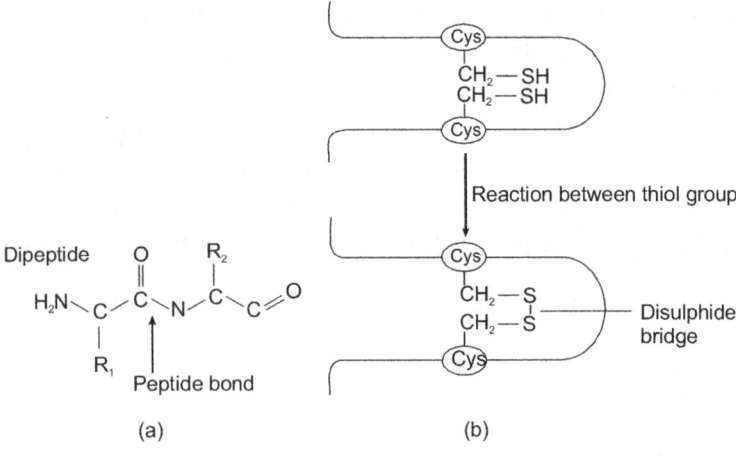

(a)   (b)

Figure 3.8   Covalent interactions in proteins (a) Peptide bond (b) Disulphide bridge

Noncovalent Interaction

Noncovalent interaction includes the following.

- Salt bridge
- Water shell
- Hydrogen bond
- van der Waals interaction
- Hydrophobic interaction

Salt bridge Salt bridge is also called **weak electrostatic bond** or **weak ionic bond** which is formed between a positively charged amino acid (arginine/lysine) and negatively charged amino acid (aspartate/glutamate) in proteins (Figure 3.9a). Salt bridge helps in stabilizing the structure of proteins.

Water shell Amino acids on the surface of proteins interact with water molecule and a thin layer of water is formed around the protein (Figure 3.9b). This layer is called water shell or hydration shell. It has been suggested that the water shell around a protein plays a part in the way a protein folds and stabilizes and thus affects the protein's function.

Hydrogen bond Hydrogen bond is formed between one of the lone pair of **electrons** on an oxygen atom and the hydrogen attached to a nitrogen atom. Hydrogen bond is involved in stabilizing the secondary structures like helix and pleated sheets (Figure 3.9c).

van der Waals interaction van der Waals interaction in protein is due to the presence of transient, electrical attraction between amino acids owing to the presence of electrical dipole. Several amino acids have quite large hydrocarbon groups in their side chains, e.g. leucine, isoleucine and phenylalanine. Temporary fluctuating dipoles in one of these groups could induce opposite dipoles in another group on a nearby folded chain. The dispersion forces set up hold the folded structure together.

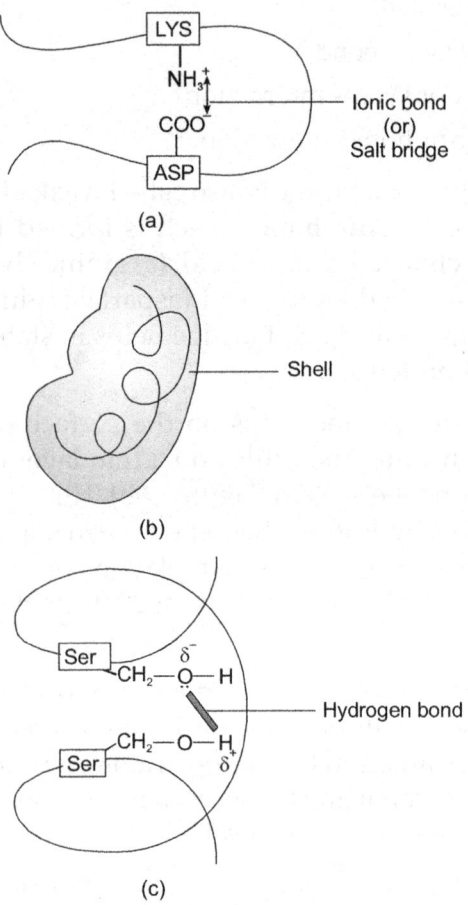

(a)

(b)

(c)

Figure 3.9   Noncovalent interactions in proteins (a) Salt bridge between amino acids in proteins (b) Water shell around protein (c) Hydrogen bond in helical protein

TYPES OF PROTEINS

Proteins can be classified in two ways: based on chemical composition and based on function.

Types of Proteins Based on Chemical Composition

Based on chemical composition proteins are of two types: simple proteins and conjugated proteins. Simple proteins are made up of amino acids only. Example, albumin and globulin. Conjugated proteins are made up of a nonprotein part (called prosthetic group) attached to a simple protein. Based on the prosthetic group, conjugated proteins are subdivided into four types:

- Glycoprotein with carbohydrates as prosthetic group, e.g. immunoglobulin.
- Lipoprotein with lipids as prosthetic group, e.g. High-density lipoprotein (HDL).
- Nucleoprotein with nucleic acid as prosthetic group, e.g. Spliceosome, histone.
- Metalloprotein with metals as prosthetic group, e.g. catalase has iron, glutathione peroxidase has selenium.

Types of Proteins Based on Biological Function

Based on biological function, proteins can be divided into 8 types:

- *Enzymes* Proteins which catalyses biochemical reactions, e.g. urease.
- *Transport proteins* Proteins which transport biological materials, e.g. haemoglobin transports

oxygen, lipoproteins transport lipids, transferrin transports iron.

- *Storage proteins*　Proteins used to store biological compounds, e.g. ferritin is the storage form of iron.

- *Contractile proteins*　Proteins which perform mechanical work like muscle contraction, e.g. actin and myosin.

- *Structural proteins*　Proteins associated with biological structures, e.g. collagen in tendon and cartilage, elastin in ligaments and keratin in hair and nails.

- *Defence proteins*　Proteins which provide defence by invading bacteria and other foreign organism, e.g. immunoglobulin.

- *Regulatory proteins or hormones*　Proteins which regulate metabolic processes, e.g. insulin regulates blood glucose.

- *Nutrient proteins*　Proteins which provide nutrition, e.g. ovalbumin (egg protein) and casein (milk protein).

PROTEIN STRUCTURE DETERMINATION

Determination of Primary Structure (Determination of Amino Acid Composition)

The steps involved in the determination of the primary structure of proteins are summarized in Figure 3.10.

Separation of amino acids　The protein sample is treated with 6N HCl at 110°C for one day. This results in the liberation of free amino acids from the protein.

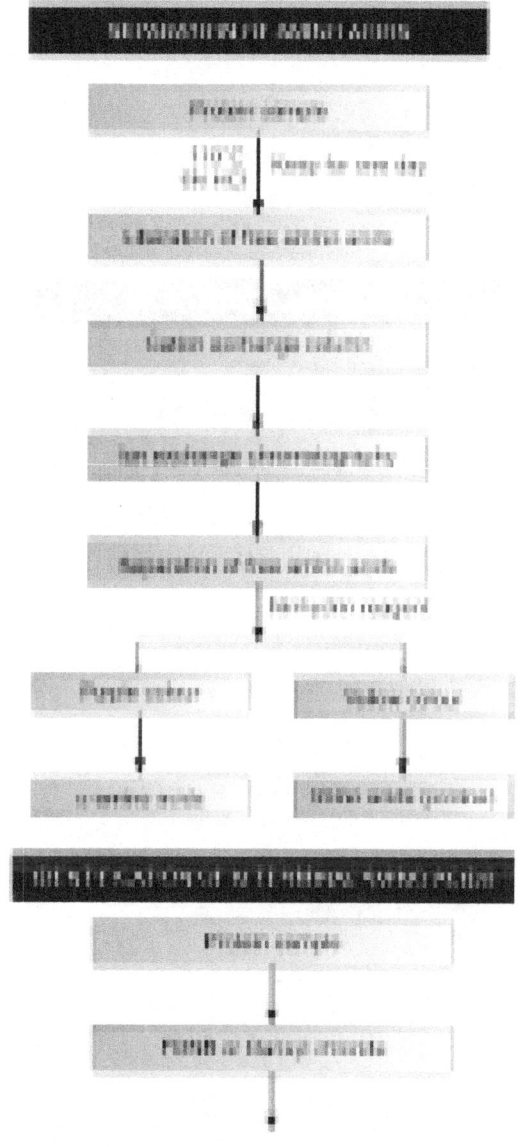

Figure 3.10 Steps involved in the determination of primary structure of proteins/peptides (*Continues*)

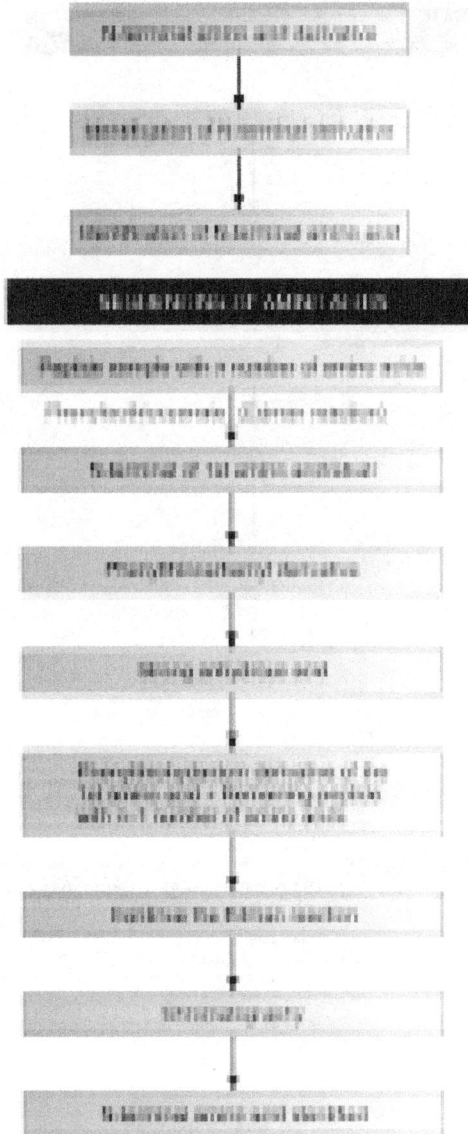

Figure 3.10 Steps involved in the determination of primary structure of proteins/peptides

The free amino acid mixture is subjected to ion-exchange chromatography using sulphonated polystyrene (cation-exchange) column. Free amino acids are separated and detected by ninhydrin reaction. α-amino acids give purple colour while imino acids (proline) give yellow colour.

Identification of N-terminal amino acids

N-terminal amino acid residue is identified using one of the N-terminal residue reagents like fluorodinitrobenzene (FDNB) or dansyl chloride. FDNB is covalently linked to the N-terminal residue making the residue resistant to acid hydrolysis.

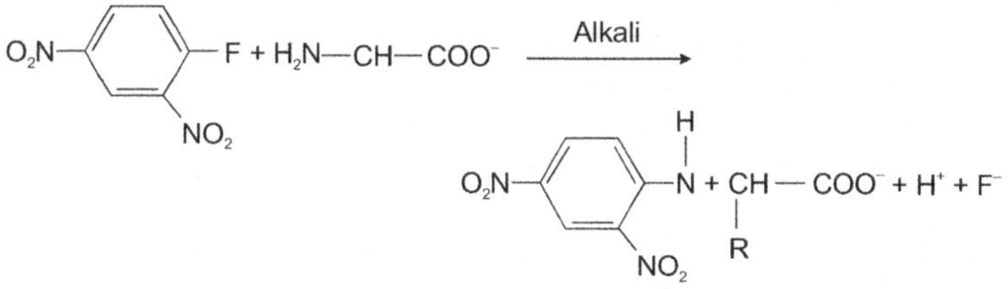

Dansyl chloride forms a fluorescent N-terminal derivative which is easily **detectable**.

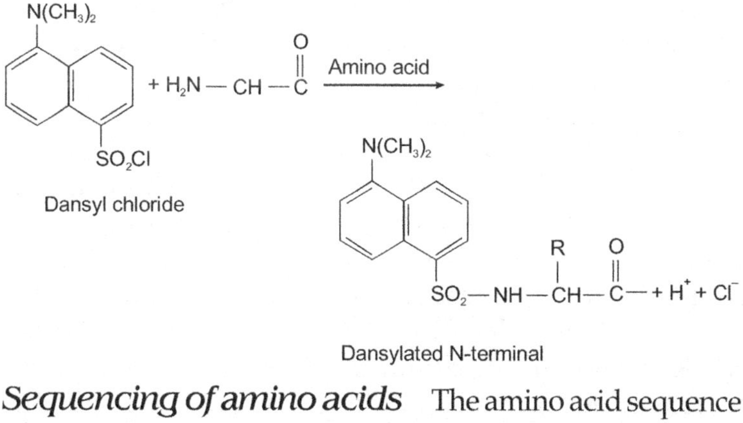

Dansyl chloride

Dansylated N-terminal

Sequencing of amino acids

The amino acid sequence of the peptide sample can be identified by **Edman reaction** using phenylisothiocyanate. Under alkaline condition, phenylisothiocyanate reacts with N-terminal group to form phenylthiocarbamyl derivative of the peptide. Phenylthiohydantoin derivative of the first amino acid can then be separated from the peptide by a strong anhydrous acid. The N-terminal amino acid can then be identified by chromatography. The remaining peptide remains intact with a new N-terminal residue. The remaining can again be treated with Edman reaction and the next amino acid can be sequenced similarly.

Determination of Secondary and Tertiary Structure

Secondary and tertiary structure of proteins can be determined using various automated techniques like:

- X-ray crystallography
- Nuclear magnetic resonance spectroscopy (NMR)
- Magnetic resonance imaging (MRI)

For X-ray diffraction, protein samples are first crystallized and beamed with X-rays. Various atoms in the protein molecules diffract X-rays differently giving unique X-ray diffraction pattern. Diffraction pattern can help determining the structure of protein crystals. NMR, in conjunction with mass spectroscopy, is often the only method for determining structures, conformations and the nature of interactions between proteins and other molecules. NMR is used to determine the three-dimensional structure of proteins at atomic resolution. Highly relevant molecular structure in solution can be obtained. Protein denaturation can also be studied using NMR. In addition, dynamic features of molecular structure, and structural, thermodynamic and kinetic aspects of interaction between proteins and other macromolecules can be studied. MRI is used to study about proteins deep inside the tissues.

PHARMACEUTICAL PROTEINS

Proteins which are produced by the process of genetic engineering and used for treatment of disease are called pharmaceutical proteins. In genetic engineering, the DNA segment coding for a specific protein is expressed in a host where it is transcribed and translated into that specific protein. Such proteins can either be secreted outside the host cell or retained within the host cell. In either case the protein should be recovered for further use.

Steps in Producing Pharmaceutical Proteins by Genetic Engineering

- Insertion of target gene into a vector/vehicle gene
- Ligation of target gene and vector
- Insertion of target gene into a suitable host (Gene transfer)
- Target gene expression and identification of recombinant host
- Recovery of target protein
- Purification and characterization of target protein

These steps are diagrammatically illustrated in Figure 3.11.

Identification and isolation of target gene The DNA segment coding for a specific protein of interest is first isolated by cutting the segment using an enzyme called **restriction endonuclease** (RE). RE is usually obtained from bacteria. For example, *Eco*R1 is a RE obtained from the bacteria *Escherichia coli*. The segmented gene is thereafter called as target gene.

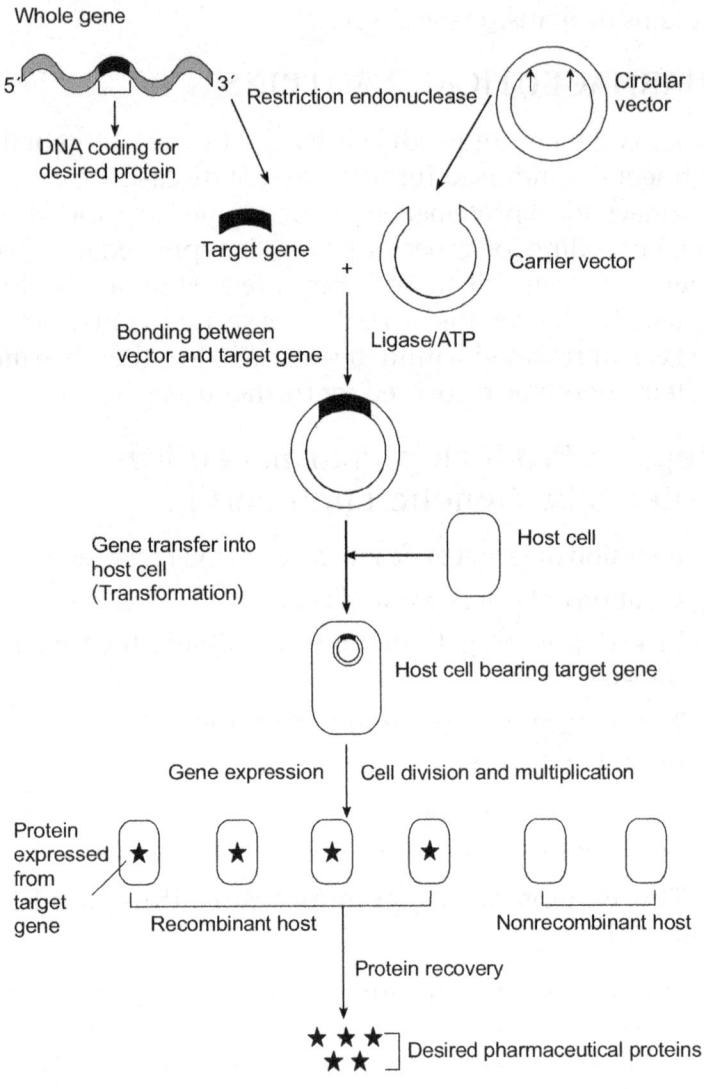

Figure 3.11   Production of pharmaceutical proteins by genetic engineering

Insertion of target gene into a vector/vehicle The target gene is inserted into a vector. The vector is also a segment of DNA used to carry the target gene. The **plasmid** (circular chromosome of bacteria) is the commonly used vector. The plasmid is linearized by cutting it with RE. Consequently, the two ends of the target gene are attached to the two ends of the linearized plasmid thus reforming a circular structure again.

Ligation of target gene and vector The ends of the target gene are ligated to the ends of vector by the formation of **phosphodiester bond**. This process is an enzyme-catalysed energy-dependent process. It is catalysed by an enzyme called **ligase** in the presence of **ATP**. Thus ligation results in a stable circular clone.

Insertion of target gene into a suitable host (Gene transfer) The vector-borne target gene is introduced into a host cell in order to be expressed (transcribed and translated). This process is called **gene transfer**. Prokaryotic hosts like bacteria can be used in general. Sometimes eukaryotic host cells like yeast cell, animal cell, plant cell or insect cell can also be used. Gene transfer into a prokaryotic host cell is called **transformation**. Gene transfer into a eukaryotic host is called **transfection**. Gene transfer can be done using a variety of methods including:

- Electroporation
- Shot gun method
- Viral transfer
- Scrap loading
- Liposome-mediated transfer.

Of these methods, electroporation and shot gun method are the more widely used methods and are reliable.

Target gene expression and identification of recombinant host   After insertion into the host cell, the target gene is duplicated and multipled by cell division and expressed into functional proteins. Genes cannot be expressed in all the host cells. Cells bearing the functional target proteins are called **recombinant host cells** and those which are devoid of the target proteins are called non-recombinant host cells. Recombinant hosts can be isolated and multiplied again to get more copies of the target gene.

Recovery of target protein   Proteins expressed in the host cell are of two types: 1) **secretory proteins** or **extracellular proteins** which are liberated into the medium of the host cell and 2) **cytosolic proteins** or **intracellular proteins** which are retained within the host cell. Secretory proteins can be recovered by **filtration** or by **centrifugation** of the culture medium. Intracellular proteins are recovered by **cell disruption** in which cell architecture is disorganized in the presence of buffer in order to release the cellular content and form crude extract. Cell disruption can be done by using blenders, homogenizers, mortar and pestle, dynomill or osmotic shock.

Purification of target protein   The target protein which is recovered from the cell should be purified and crystallized for pharmaceutical and medical use. Purification can be done by ion-exchange chromatography gel filtration chromatography, high pressure liquid chromatography (HPLC) or affinity chromatography. Purified proteins can be characterized for further study. Characterization can be done by using techniques like analytical ultracentrifugation, isoelectric focusing, gel filtration chromatography, SDS-PAGE and mass spectroscopy.

Examples for Pharmaceutical Proteins

Various therapeutic enzymes, hormones, interleukins, interferons clotting factors and growth factors of pharmaceutical importance are produced by recombinant DNA technology. These are listed in Table 3.4.

Table 3.4 Diseases cured by recombinant pharmaceutical proteins

Proteins	Disease cured
Enzymes	
L-asparaginase	Cancer
Superoxide dismutase	Oxygen toxicity
Urokinase	Heart attack
DNase	Cystic fibrosis
Glucocerebrosidase	Gaucher's disease
Galactosidase	Inherited β-galactosidase deficiency
Penicillinase	Penicillin allergy
Papain	Digestive disorder
Hormones	
Insulin	Diabetes
Calcitonin	Osteomalacia
Somatotropin	Growth defect
Growth hormone	Growth defect
Erythropoietin	Anaemia
ACTH (Adrenocorticotropic hormone)	Rheumatic disease
Chorionic gonadotropin	Infertility

(Contd.)

Table 3.4 (Continued)

Proteins	Disease cured
Interleukin and interferons	
Interleukin I	Asthma
Interleukins	Cancer
Interferons	Cancer, immune disorders
Clotting factor	
Factor VIII	Haemophilia A
Factor IX	Haemophilia B
Growth factor	
Platelet-derived growth factor	Atherosclerosis
Epidermal growth factor	Burns
Neural growth factor	Brain damage

METABOLISM OF AMINO ACIDS

Catabolism or Degradation of Amino Acids

Amino acids are degraded in two ways. First, the amino group is degraded to **ammonia** via **transamination** and **oxidative deamination** reactions (Figure 3.12). Transamination is catalysed by transaminase and oxidative deamination is catalysed by dehydrogenase. Ammonia is converted to urea (Figure 3.13) via **urea cycle** and excreted in the urine.

Second, the carbon skeleton is degraded. Depending upon the end product of degradation of the carbon skeleton, amino acids are classified as **ketogenic amino acids** and **glucogenic amino acids**.

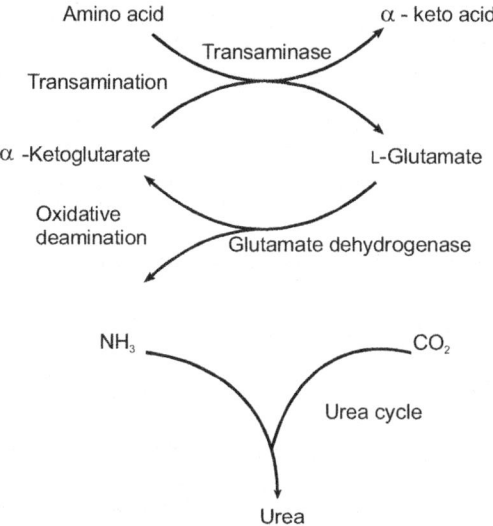

Figure 3.12 Degradation of amino acids via transamination and oxidative deamination

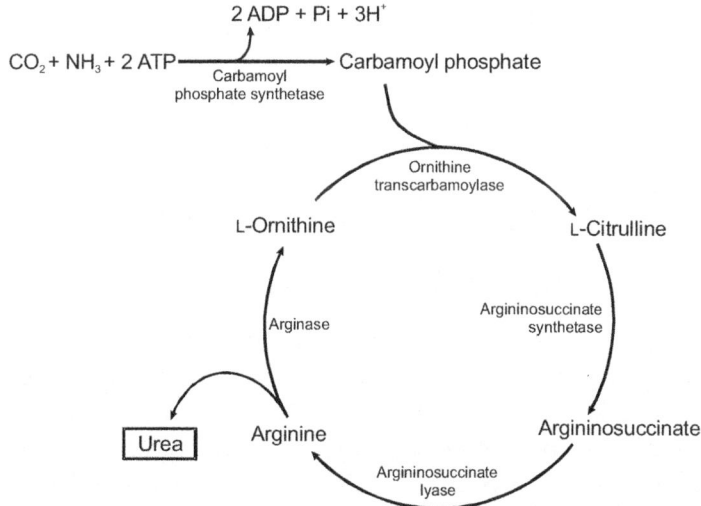

Figure 3.13 Urea cycle (Degradation of ammonia)

Isoleucine, leucine and tryptophan are degraded to acetyl-CoA which is then converted to acetoacetyl-CoA. Lysine, phenylalanine, tyrosine and tryptophan are degraded directly into acetoacetyl-CoA. Acetoacetyl -CoA is finally converted to ketone bodies (Figure 3.14). Hence these amino acids are ketogenic amino acids. Alanine, glycine, cysteine, serine, threonine, asparagine, aspartate, phenylalanine, methionine, valine, glutamate, glutamine, histidine, proline and arginine are degraded to TCA cycle intermediates finally forming glucose (Figure 3.15). These amino acids are glucogenic amino acids.

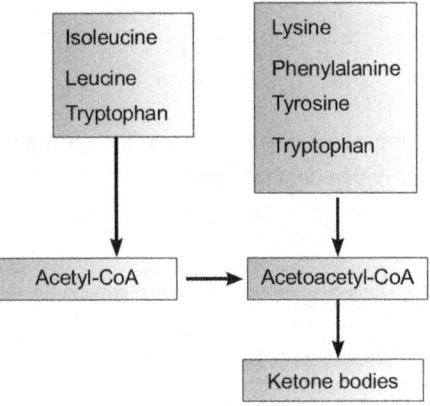

Figure 3.14   Degradation of ketogenic amino acids to ketone bodies

Biosynthesis of Amino Acids

Essential amino acids **cannot** be synthesized in the body. Hence they should be supplied in the diet. Non-essential amino acids are synthesized in the body from various **precursors** in the presence of various enzymes and coenzymes (Figure 3.16). Alanine is synthesized from pyruvate. Aspartate and asparagine are synthesized from oxaloacetate.

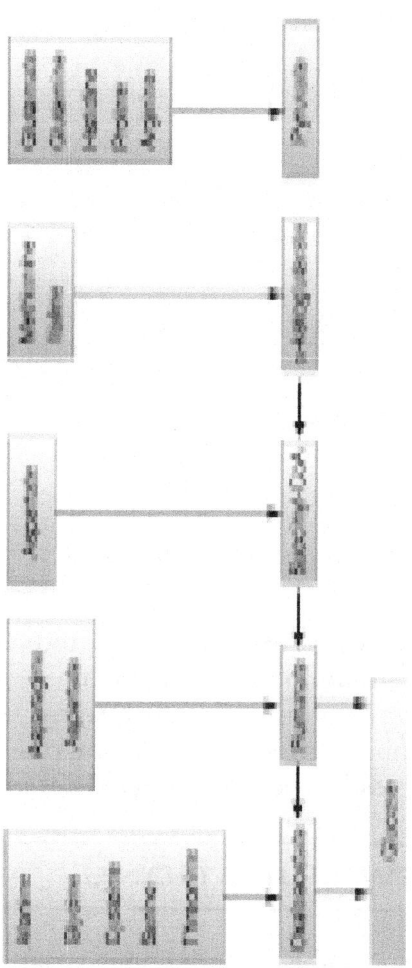

Figure 3.15   Degradation of glucogenic amino acids to glucose

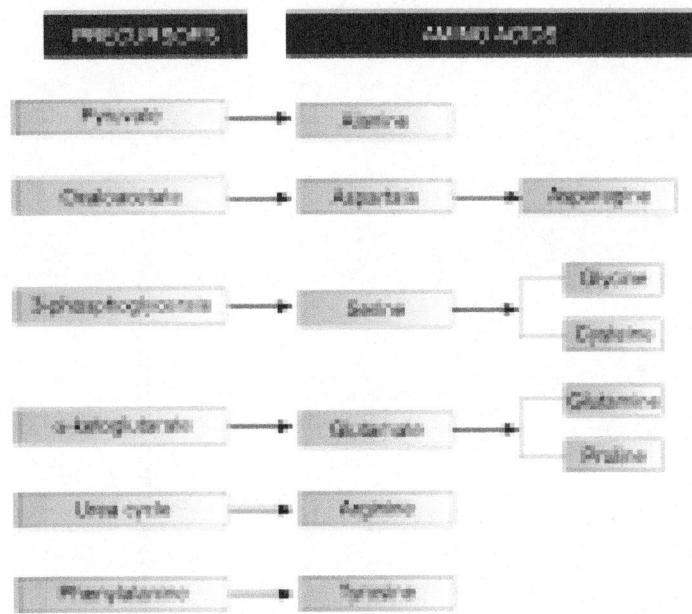

Figure 3.16 Biosynthesis of non-essential amino acids from their precursors

Serine, glycine and cysteine are obtained from 3-phosphoglycerate. Glutamine and proline are derived from glutamate which in turn is synthesized from α-ketoglutarate. Arginine is produced as an intermediate of urea cycle. Tyrosine is obtained from phenylalanine.

REVIEW QUESTIONS

1. Give the general structural formula for amino acid.
2. Define zwitterions.
3. Amino acids are amphoteric—Explain.
4. What is ninhydrin reaction?

5. How do amino acids react in Sangers reaction?
6. What is xanthoproteic test?
7. What is disulphide bond? What is its significance in protein structure?
8. What is hydrogen bond?
9. Define glycoprotein and metalloprotein with an example.
10. What are transport proteins? Give an example.
11. What are contractile proteins? Give an example.

CHOOSE THE CORRECT ANSWER

1. Sulphur is present in
   - (a) lysine
   - (b) cysteine
   - (c) arginine
   - (d) phenylalanine
2. Phenylalanine is a
   - (a) aliphatic amino acid
   - (b) aromatic amino acid
   - (c) acidic amino acid
   - (d) imino acid
3. Proline gives yellow colour with
   - (a) Sanger's reagent
   - (b) FDNB reagent
   - (c) Ninhydrin reagent
   - (d) All the above
4. Sakaguchi's test is specific for
   - (a) alanine
   - (b) histidine
   - (c) arginine
   - (d) aspartate
5. β-meander is a
   - (a) primary structure
   - (b) secondary structure
   - (c) supersecondary structure
   - (d) tertiary structure

JENS C. SKOU

Discovered "Na$^+$ K$^+$ ATPase", an ion-transporting enzyme.

Enzymes are catalytic proteins. They are colloidal, high molecular weight, non dialysable, denaturable structurally diverse group of proteins. Biochemical reactions taking place within the living cells are catalysed by enzymes (hence the name biocatalysts). Enzymes increase the rate of reaction without themselves undergoing any change and without altering the equilibrium of the reaction.

The speed with which the enzyme catalyses a reaction is called as **velocity** or **rate** of that reaction. It is denoted as V. The substance on which the enzyme acts is called **substrate** (S) and the new substance that is generated after the reaction is called **product** (P).

RIBOZYMES AND ISOZYMES

Strictly speaking, all enzymes are not proteins because few RNA molecules have catalytic properties and are called ribozymes. They are used in the catalysis of RNA splicing, in the maturation of tRNA and in the translation process (peptide bond formation). **RNAse P,** and RNA component of peptidyl transferase are examples for ribozymes.

Sometimes a single enzyme can exist in more than one form. They are called isoforms or isozymes or isoenzymes. They catalyse the same reaction but differ in their amino acid sequence and electrophoretic mobility. Lactate dehydrogenase (LDH) is a classic example for isozymes. **LDH** exists in five forms denoted by LDH1 to LDH5. Other examples are **hexokinase** which exists in four forms type 1 to type 4 and **angiotensin-converting enzyme** which exists in two forms, somatic form and germinal form. Isoenzymes are used in the diagnosis of diseases related to a specific tissue.

ENZYME ACTIVITY

The amount of enzyme that catalyses the formation of one micromole of product in one minute under optimal conditions of pH, temperature and ionic strength is called One IU of an enzyme.

The amount of enzyme catalysing the conversion of one mole of substrate to product in one second is called one Katal. One Katal = 6×10^7 IU.

PROPERTIES OF ENZYMES

Enzymes possess the following distinctive properties.

1. *Catalytic power*   The ratio of **the rate of catalysed reaction to the rate of uncatalysed reaction** is called catalytic power of an enzyme. For example, the rate of hydration of carbon dioxide to form bicarbonate is 1.3×10^{-1} per second in the absence of enzyme. But the rate of the same reaction in the presence of the enzyme carbonic anhydrase is 1×10^6. Thus the catalytic power of carbonic anhydrase is 7.7×10^7.

PROPERTIES OF ENZYME
Catalytic power
Specificity
Milder reaction condition
Reversibility
Denaturation

2. *Specificity* The **extraordinary ability** of the enzyme to recognize a specific substrate to catalyse a specific reaction is called the specificity of the enzyme for example, urease is specific for urea and hence it reacts only with urea to form ammonia.

$$Urea \xrightarrow{\text{Urease}} Ammonia + Carbon\ dioxide$$

3. *Milder reaction condition* Enzyme catalysis occurs under milder conditions of temperature, atmospheric pressure and nearly neutral pH. But, there are certain unique groups of enzymes called extremozymes which can withstand extreme environmental conditions like very high temperature, very low temperature, very high saline concentration and desiccation. These enzymes are derived from a group of bacteria called extremophiles. For example, *Taq* polymerase which is resistant to a temperature of about 100°C is derived from a bacteria called *Thermus aquaticus*. This enzyme is used in polymerase chain reaction (PCR).

4. *Reversibility* Enzymes can catalyse reactions in both directions depending upon the availability of suitable energy source and solvent conditions.

5. *Denaturation* Denaturation refers to change in conformational changes of proteins that leads to inactivation of enzymes. Enzymes are denatured by high temperature, strong acids and strong alkali.

FACTORS INFLUENCING ENZYME ACTIVITY

Various factors influence the activity of enzymes. These factors include substrate concentration, reaction temperature and pH of the buffer or reaction medium.

Effect of Substrate Concentration

Effect of substrate concentration on enzyme activity can be well described by plotting the velocity of the reaction against substrate concentration as shown in Figure 4.1.

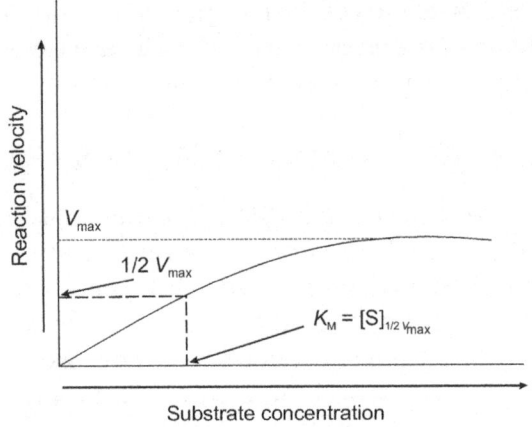

Figure 4.1   Effect of substrate concentration on enzyme activity

At the initial stage of the reaction, the velocity of the enzyme-catalysed reaction increases **linearly** with an increase in substrate concentration (shown as linear region). This is also termed **initial velocity** of the reaction. Generally it is assumed that the velocity of the enzyme-catalysed reaction is directly proportional to the enzyme concentration in a reaction mixture. As the substrate concentration is raised gradually, the velocity reaches a maximum (V_{max}). After V_{max}, the velocity is **constant** even with increase in substrate concentration. This leads to a **plateau** region. Thus a hyperbola is obtained.

Effect of Temperature

Effect of temperature on the rate of enzyme-catalysed reaction is represented as plot of velocity vs. temperature (Figure 4.2).

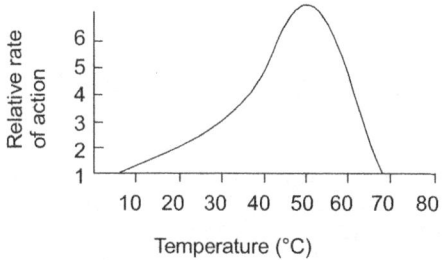

Figure 4.2   Effect of temperature on rate of enzyme-catalysed reaction

The rate of the enzyme-catalysed reaction increases with initial rise in temperature until the optimum temperature. This rise is due to the effective collision between reactive groups of substrate and enzyme. At optimum temperature the velocity is maximum. Further increase in temperature leads to a fall in velocity due to disturbance in conformation of the active site of the enzyme. This leads to a bell-shaped curve.

Effect of pH

The effect of pH on the velocity of an enzyme-catalysed reaction is represented in Figure 4.3. The **bell-shaped** curve of pH vs. V is similar to that of T vs. V. In the initial stage of the reaction, the velocity increases with increase in pH due to increase in enzyme–substrate **binding**. At a particular pH called the **optimum pH** (usually 6–8), the velocity reaches a maximum. Beyond the optimum pH, the 3D structure of the enzyme is altered leading to a **dissociation** of the E–S complex and a fall in velocity.

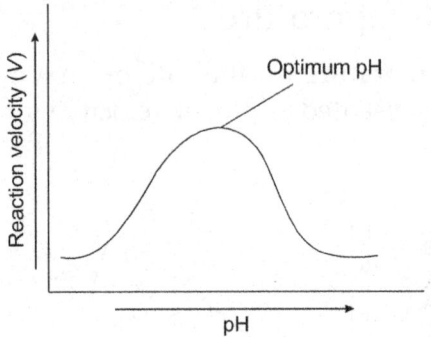

Figure 4.3   Effect of pH on velocity of enzyme-catalysed reaction

STRUCTURE OF ENZYMES

Enzymes have **properly folded** active conformation. The part of the enzyme responsible for catalysis is called **active site.** The active site is functionally divided into two subsites (Figure 4.4) — **Specificity subsite (substrate binding site)** which is the part of the active site where recognition of the substrate takes place and **reaction subsite** which is the part of the active site where chemistry occurs.

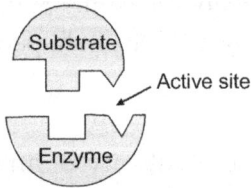

Figure 4.4   Structure of enzyme

The substrate (S) binds to the active site of the enzyme (E) forming the ES complex. The ES complex is converted into a **high energy** state called **transition state** complex (ES‡). Subsequently, the transition state is converted into product (P) and the free enzyme (E) is released.

$$E + S \rightarrow ES \rightarrow ES^{\ddagger} \rightarrow P + E$$

MODELS FOR ENZYME–SUBSTRATE REACTION

Enzyme–substrate binding can conventionally be explained by two models:

Fischer's Lock-and-Key Model

According to this model the active site of the enzyme is rigid and **predetermined** to be complementary to the substrate. Thus, the substrate fits into the active site as a key fits into the lock (Figure 4.5a).

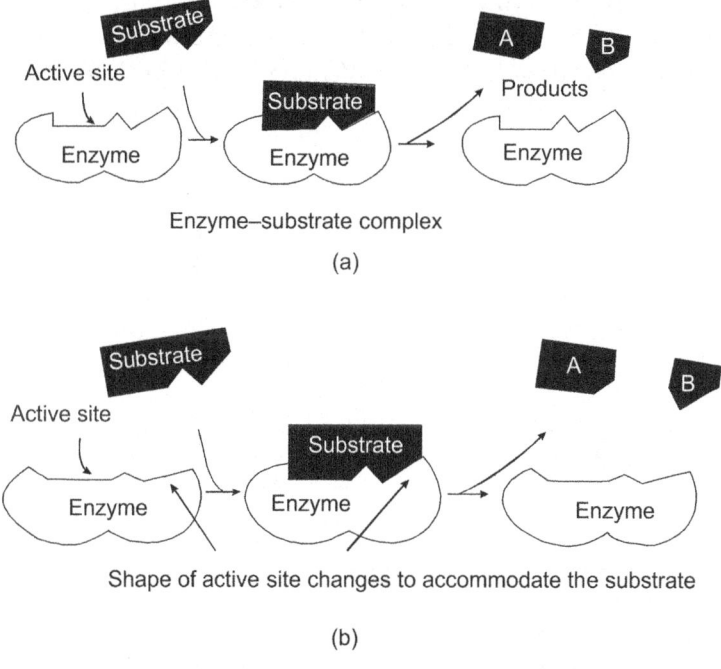

Figure 4.5 Two models for enzyme–substrate binding (a) Fischer's lock-and-key model (b) Koshland's induced fit model

Koshland's Induced Fit Model

In this model, the active site of the enzyme is flexible and changes shape on substrate binding, i.e., the complementarity is induced after the binding of the substrate to the enzyme (Figure 4.5b).

MECHANISM OF ENZYME ACTION

Energy required for the formation of transition state is called **activation energy** (ΔG^*). Generally, an enzyme increases the reaction rate by **lowering** the activation energy. This can be explained as follows: After the substrate fits into the active site of the enzyme as per the model described above, free energy is released and it is called **intrinsic binding energy** and is denoted as ΔGb. When ES complex is formed, the entropy of the S is lost.

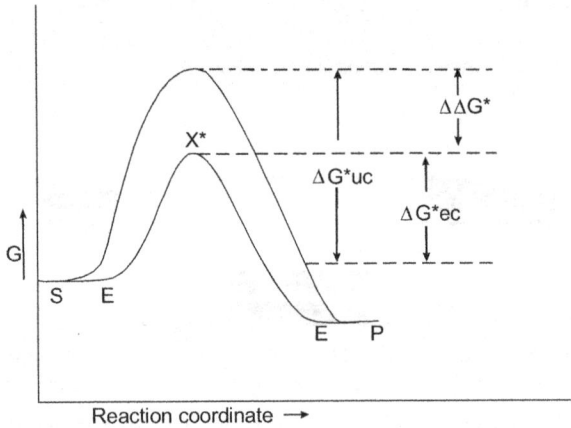

G = Free energy

ΔG^* uc = Activation energy for uncatalysed reaction

ΔG^* ec = Activation energy for enzyme-catalysed reaction

$\Delta \Delta G^*$ = Amount of activation energy lowered by enzyme (catalytic efficiency).

Figure 4.6　Transition state diagram showing mechanism of enzyme action

The entropy loss is denoted as $-T\Delta S$. The amount by which the activation energy is lowered by the enzyme is called **catalytic efficiency** and is denoted as $\Delta\Delta G^*$. ΔGb acts as a major source of free energy for the enzyme to lower the activation energy. ΔGb is equal to the amount by which ΔG^* is lowered, i.e., $\Delta Gb = \Delta G^*$. This catalytic concept could be described by a **transition state diagram** (or reaction coordinate diagram) which is a plot of free energy (G) vs. progress of the reaction (reaction coordinate) (Figure 4.6).

RATE EQUATION FOR ENZYME-CATALYSED REACTION—MICHAELIS –MENTEN EQUATION

The substrate concentration at which the velocity of the enzyme-catalysed reaction is equal to half of the maximum velocity (V_{max}) is called K_m value (Figure 4.7). Four factors are involved in an enzyme-catalysed reaction namely substrate concentration ([S]), velocity (V), maximum velocity (V_{max}) and K_m.

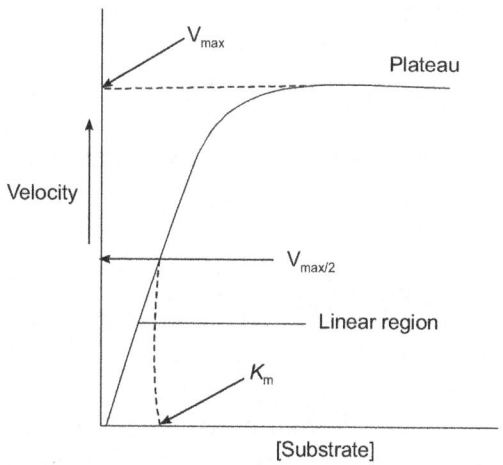

Figure 4.7 Course of an enzyme-catalysed reaction

These parameters can be related by an equation called rate equation or velocity equation or Michaelis–Menten equation:

$$V = \frac{V_{max}[S]}{K_m + [S]}$$

This equation was proposed by Michaelis and Menten and hence the name Michaelis–Menten equation or MM equation.

ENZYME INHIBITION

Inhibition or suppression of cellular enzymatic activity by low molecular weight compounds like drugs, antibiotics, poisons and certain metabolites is called enzyme inhibition. Molecules that inhibit the activity are called inhibitors. Enzyme inhibition is of three types – competitive inhibition, non-competitive inhibition and uncompetitive inhibition. These three types of inhibition are represented in Figure 4.8. Examples for these three types of inhibition are shown in Table 4.1.

In competitive inhibition, the substrate and the inhibitor exhibit **structural similarity** hence, they compete for the same active site. For example, succinate dehydrogenase (SDH) is competitively inhibited by malonate which has structural similarity with succinate, the substrate of the reaction.

$$\text{Succinate} \xrightarrow{\text{SDH}} \text{Fumarate}$$

$$\uparrow \ominus$$

Malonate

In non-competitive inhibition the substrate and the inhibitor have **no structural similarity** and hence do not compete for the same active site. Non-competitive inhibitors bind to the **allosteric site** (site other than the active site) of the enzyme and alter the three-dimensional structure of the active site so that the substrate cannot fit into the active site. Non-competitive inhibitor binds either to the free enzyme or to the enzyme–substrate complex. For example, heavy metals interact with the thiol group of the enzyme forming mercaptide complex which is inactive.

$$E-SH + Hg^{2+} \rightarrow E-S-Hg + H^{+}$$

Inactive
mercaptide
complex

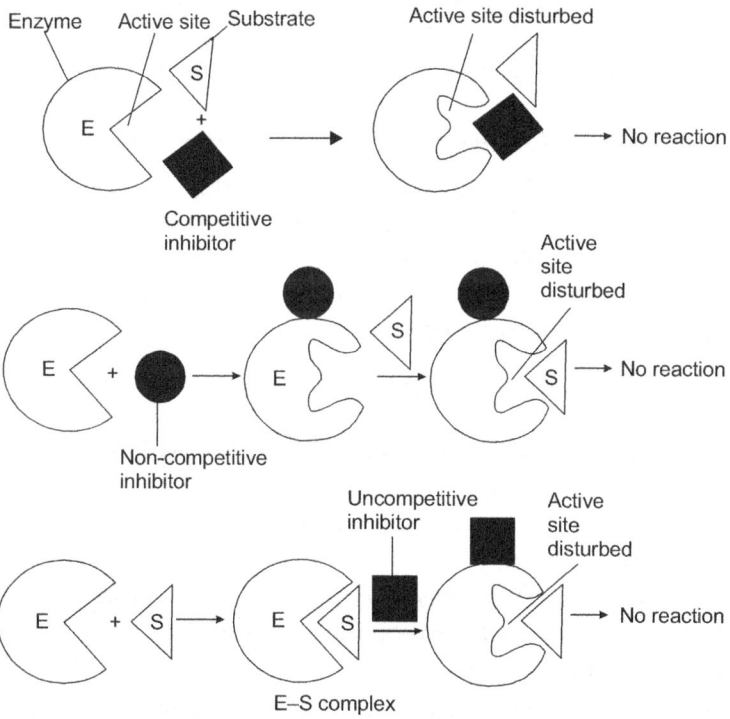

Figure 4.8 Types of enzyme inhibition

Table 4.1   Types of enzyme inhibition

Inhibition	Example	Inhibitor
Competitive	Succinate dehydrogenase	Malonate
	Dihydrofolate reductase	Methotrexate
	Xanthine oxidase	Allopurinol
Non-competitive	Porphobilinogen synthase	Iron
	Ferrochelatase	Iron
	Aldolase	Fluoride
	Catalase	Cyanide
Uncompetitive	Alkaline phosphatase	L-phenylalanine

In uncompetitive inhibition, the inhibitor has **no affinity** for substrates. Inhibitors bind to the allosteric site of the **enzyme–substrate complex** only but not to the free enzyme. For example, L-**phenylalanine** uncompetitively inhibits intestinal alkaline phosphatase.

COENZYMES AND COFACTORS

Coenzymes are low molecular weight **non-protein organic** molecules that are loosely bound to enzymes and transfer chemical groups between enzymes. Coenzymes are sometimes referred to as **cosubstrates**. Coenzymes are carriers of specific **functional groups** during biochemical reactions. Based on the precursor molecule, coenzymes are classified into two types:

1. Vitamin-derived coenzymes, e.g. NADH, FADH, coenzyme A (CoA) and thymine pyrophosphate. NAD is the coenzyme for lactate dehydrogenase.

2. Non-vitamin-derived coenzymes, e.g. biopterin, coenzyme Q (CoQ) and ATP.

Transition-state metal ions which enhance the reaction rate by binding to various groups (ligands) of enzymes are called **cofactors.** Iron, molybdenum, copper, zinc, manganese, magnesium, nickel, selenium, fluorine and vanadium are significant cofactors involved in many biochemical reactions. For example, zinc is the cofactor for the enzyme carbonic anhydrase. Catalase depends on iron as cofactor. Selenium is the cofactor for glutathione peroxidase.

APPLICATIONS OF ENZYMES

Enzymes are used in the following ways.

- As diagnostic agents, for example, lactate dehydrogenase is used in the diagnosis of heart disease. Pepsin is used in the diagnosis of gastric mucosal damage.

- As analytical agents, for example, hexokinase is used to estimate glucose. Urease is used to estimate urea.

- As therapeutic agents, for example, pepsin is used in treating digestive disorders and penicillinase is used in treating penicillin allergy. DNase is used for the treatment of cystic fibrosis.

- In recombinant DNA technology and cloning, for example, restriction endonuclease is used to cut DNA and ligase is used to join two DNA segments by phosphodiester bond.

- In industrial processes, protease is used in beverage industry, amylase is used in confectioneries, and in the detergent, paper and textile industries.

REVIEW QUESTIONS

1. Define enzymes.
2. What is international unit (IU) of an enzyme?
3. What is Katal?
4. What do you mean by catalytic power of an enzyme?
5. What is meant by specificity of an enzyme?
6. What is denaturation?
7. What is activation energy?
8. Write the rate equation for enzyme-catalysed reaction.
9. Write about competitive inhibition and give an example.
10. What is non-competitive inhibition? Give an example.
11. What is an uncompetitive inhibitor?
12. List out the applications of enzymes.

CHOOSE THE CORRECT ANSWER

1. Michaelis and Menten proposed
   (a) enzyme mechanism   (b) enzyme inhibition
   (c) enzyme kinetics   (d) enzyme units
2. An example for ribozyme is
   (a) DNase P   (b) RNase P
   (c) DNase H   (d) RNase H
3. A classic example for isozyme is
   (a) lactate dehydrogenase   (b) DNA polymerase
   (c) oxidoreductase   (d) ligase

4. Fischer proposed
   (a) lock-and-key model   (b) induced fit model
   (c) enzyme kinetics   (d) enzyme inhibition model

5. Gastric mucosal damage can be diagnosed by measuring the activity of
   (a) ribozyme   (b) pepsin
   (c) chymotrypsin   (d) trypsin

LUIS F. LELOIR

Discovered sugar nucleotides and their role in the biosynthesis of carbohydrates.

Carbohydrates are hydrated carbon molecules with the general formula $C_nH_{2n}O_n$ where n = number of carbon atoms. For example $C_6H_{12}O_6$ is glucose. It has 6 carbon and 6 oxygen atoms (n value) and 12 hydrogen atoms ($2n$ value). Carbohydrates can exist in two forms **polyhydroxyketones** or **polyhydroxy aldehydes.**

FUNCTIONS OF CARBOHYDRATES

Carbohydrates perform the following important functions:

- Provide the majority of energy
- Structural units of cell wall, e.g. cellulose.
- Structural units of cell membranes.
- Serve as metabolic intermediates.
- Comprise a large portion of DNA and RNA in the form of deoxyribose and ribose.

- Play a role in lubrication of joints and aid cell adhesion.
- Play a role in immunity in conjugation with protein (glycoprotein).

NOMENCLATURE OF SUGARS BASED ON NUMBER OF CARBON ATOMS

The building blocks of carbohydrates are called sugar residues (like amino acid residues are building blocks of proteins). Sugars with three carbon atoms are called trioses; those with four carbon atoms are called tetroses and those with five carbons and six carbons are called pentoses and hexoses respectively. Hexose is the most abundant sugar, e.g. glucose and fructose. Triose is the simplest sugar, e.g. glyceraldehyde. Well known examples for each type of sugar are given in Table 5.1.

Table 5.1   Examples for various types of sugars

Sugar	Number of carbon atoms present	Example
Triose	3	Glyceraldehyde
		Dihydroxyacetone phosphate
Tetrose	4	Erythrose
		Erythrulose
		Threose
Pentose	5	Xylose
		Ribose
		Arabinose
		Lyxose
Hexose	6	Glucose
		Gulose

(Contd.)

Table 5.1   (Continued)

Sugar	Number of carbon atoms present	Example
		Mannose
		Idose
		Talose
		Allose
		Altrose
Heptose	7	Sedoheptulose

CLASSIFICATION OF CARBOHYDRATES

Classification of Carbohydrates Based on Functional Groups

As mentioned earlier carbohydrates are polyhydroxyaldehydes or polyhydroxyketones. This means that they have either aldehyde group or keto group. Based on this carbohydrates are divided into two types: aldose and ketose. Aldoses have aldehyde as the functional group, e.g. glucose, galactose and maltose. Ketoses have ketone group as the functional group, e.g. fructose, ribulose and xylulose. The structures of some aldoses and ketoses are shown in Figure 5.1.

Classification of Carbohydrates Based on Number of Sugar Residues

Based on the number of sugar residues, carbohydrates can be classified into monosaccharides, disaccharides, trisaccharides and polysaccharides. Monosaccharides are made up of only one sugar residue, e.g. glucose. Disaccharides are made up of two sugar residues, e.g. maltose made up of two glucose residues, lactose made up of one glucose residue and one galactose residue, sucrose made up of one glucose and one fructose residue. Trisaccharides have three sugar residues example, raffinose and maltotriose. A short chain of carbohydrates with 4–10 sugar residues are called oligosaccharides. Carbohydrates with more than ten sugar residues are called polysaccharides; example, starch and cellulose.

STRUCTURE OF CARBOHYDRATES

Structure of carbohydrates can be learned under three headings:

1. Open chain form
2. Cyclic form
3. Ring form

Open Chain Form

Open chain form of sugar is linear with its functional group (aldehyde/ketone group) at one end and alcoholic group at the other end. For example open chain form of glucose is shown in Figure 5.2. In glucose, carbon atoms are linked one below the other in a linear fashion. The first carbon atom is linked to the aldehyde group and the sixth carbon is linked to alcoholic group ($-CH_2OH$). Carbon 2 through 5 have hydrogen on one side and hydroxyl group on the other side. 2nd, 4th and 5th carbon have hydrogen on their left and OH on the right side, whereas in the 3rd carbon, OH is on the left and hydrogen is on the right side.

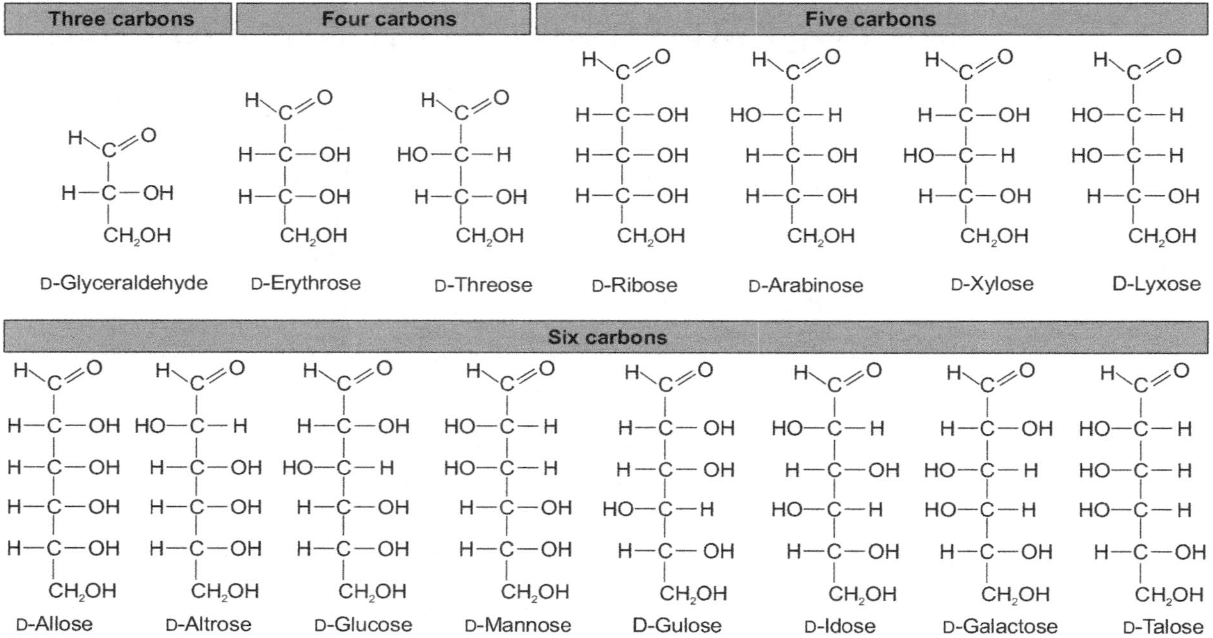

Figure 5.1 Structure of aldoses and ketoses (a) D-Aldoses (b) D-Ketoses (*Continues*)

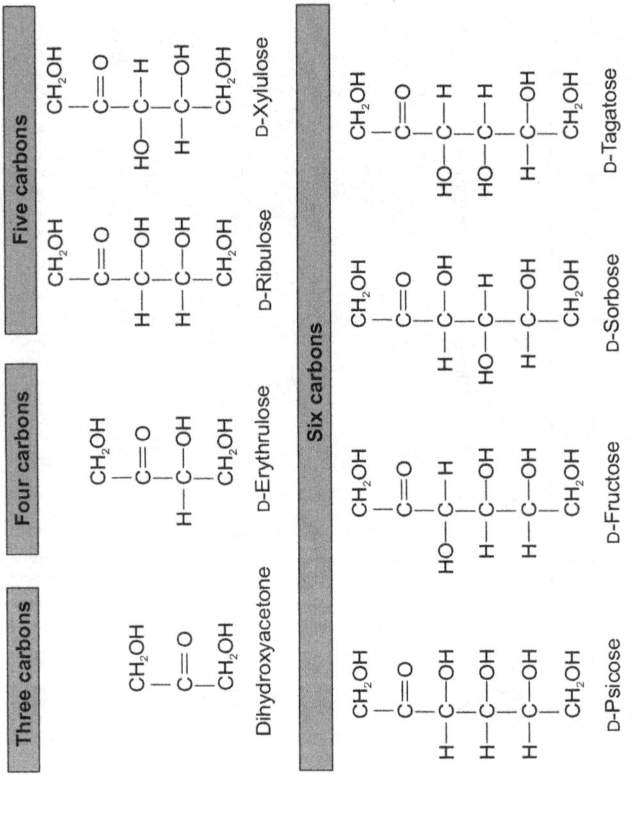

Three carbons | **Four carbons** | **Five carbons**

Six carbons

CH_2OH
$C=O$
CH_2OH
Dihydroxyacetone

CH_2OH
$C=O$
$H-C-OH$
CH_2OH
D-Erythrulose

CH_2OH
$C=O$
$H-C-OH$
$H-C-OH$
CH_2OH
D-Ribulose

CH_2OH
$C=O$
$HO-C-H$
$H-C-OH$
CH_2OH
D-Xylulose

CH_2OH
$C=O$
$H-C-OH$
$H-C-OH$
$H-C-OH$
CH_2OH
D-Psicose

CH_2OH
$C=O$
$HO-C-H$
$H-C-OH$
$H-C-OH$
CH_2OH
D-Fructose

CH_2OH
$C=O$
$H-C-OH$
$HO-C-H$
$H-C-OH$
CH_2OH
D-Sorbose

CH_2OH
$C=O$
$HO-C-H$
$HO-C-H$
$H-C-OH$
CH_2OH
D-Tagatose

(b)

Figure 5.1 Structure of aldoses and ketoses (a) D-Aldoses (b) D-Ketoses

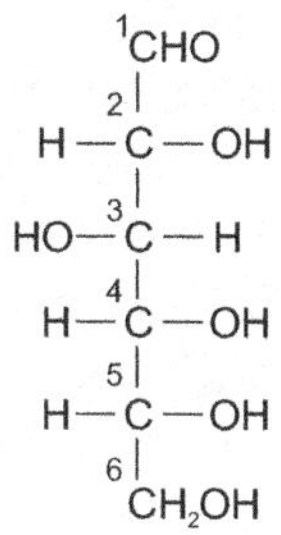

$$^1CHO$$
$$H-\overset{2}{C}-OH$$
$$HO-\overset{3}{C}-H$$
$$H-\overset{4}{C}-OH$$
$$H-\overset{5}{C}-OH$$
$$\overset{6}{C}H_2OH$$

Figure 5.2 Structure of open chain form of glucose

Stereoisomerism in monosaccharides Presence of chiral carbon gives rise to steroisomerism in monosaccharides. All monosaccharides have at least one chiral carbon (i.e., carbon with four variable groups) due to which they exist as stereoisomers. Stereoisomers may be defined as different structural forms of the same compound which differ in their spatial arrangements of groups. The possible number of stereoisomers for a monosaccharide depends on the number of chiral carbons and is given by the formula 2^n where n = number of chiral carbons in the monosaccharide. For example, glyceraldehyde has one chiral carbon and hence $2^n = 2^1 = 2$. There are two stereoisomers for glyceraldehydes. One is L-glyceraldehyde and the other is D-glyceraldehyde. Structure of L-glyceraldehyde and D-glyceraldehyde are given in Figure 5.3.

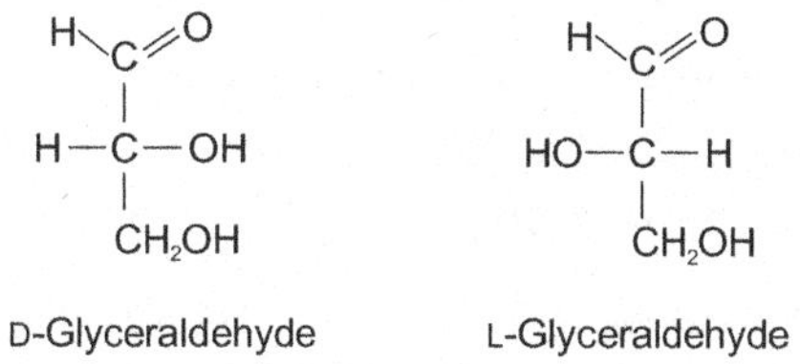

$$H-C\overset{O}{\diagup}$$
$$H-C-OH$$
$$CH_2OH$$
D-Glyceraldehyde

$$H-C\overset{O}{\diagup}$$
$$HO-C-H$$
$$CH_2OH$$
L-Glyceraldehyde

Figure 5.3 Structures of D- and L-glyceraldehydes

Other sugars are considered to be derived from glyceraldehyde. D-sugar is one that matches the configuration of D-glyceraldehyde around the asymmetric carbon that is farthest from the reactive aldehyde or ketone group. L-sugar correspondingly matches L-glyceraldehyde. In other words, D and L configuration is given based on the position of −OH group on the penultimate carbon. If the −OH group of the penultimate carbon is on the left, the sugar is L-sugar; if it is on the right, it is D-sugar. For example, L-glucose and D-glucose are shown in Figure 5.4.

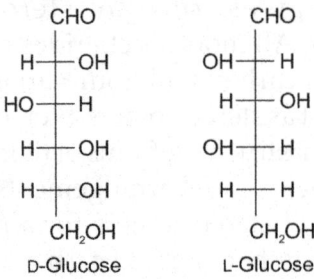

Figure 5.4   Structure of D-glucose and L-glucose

Enantiomers   Compounds that deflect plane polarized light either to the right (clockwise) or to the left (anti-clockwise) are called enantiomers or optical isomers. Enantiomers are mirror images of each other. Monosaccharides are enantiomers. Monosaccharides that deflect plane polarized light to the right (clockwise) are called dextrorotatory sugar and denoted as d(+). Sugars that deflect the light to the left (anti-clockwise) are called laevorotatory and denoted as l(−). Glucose is dextrorotatory and fructose is laevorotatory.

Epimers   Monosaccharides which differ in the position of −OH group around a single carbon are called epimers. For example, consider glucose, galactose and mannose as shown in Figure 5.5. These sugars differ by one carbon which is highlighted.

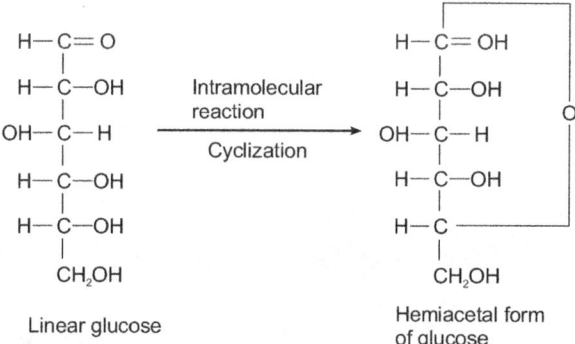

```
      CHO              CHO                CHO
       |                |                  |
  H ── OH          OH ── C ── H       OH ── H
       |                |                  |
  HO ── H          OH ── C ── H       H ── OH
       |                |                  |
  H ── OH          H ── C ── OH       OH ── H
       |                |                  |
  H ── OH          H ── C ── OH       OH ── H
       |                |                  |
    CH₂OH            CH₂OH              CH₂OH

  D-Glucose        D-Mannose          L-Glucose
```

Figure 5.5 Epimers (Glucose, mannose and galactose differ in third carbon atom)

Cyclic Form

Cyclic form of sugar is formed due to cyclization and bond formation between different carbon atoms of that sugar residue. For example, in glucose a bond is formed between the carbonyl group of the first carbon and hydroxyl group of the fifth carbon resulting in cyclization of the structure. The cyclic form is called hemiacetal form (Figure 5.6).

```
   H ── C = O                              H ── C = OH ─┐
        |                                       |       │
   H ── C ── OH      Intramolecular        H ── C ── OH  │
        |            reaction                   |        │ O
  OH ── C ── H       ──────────────>      OH ── C ── H   │
        |            Cyclization               |         │
   H ── C ── OH                            H ── C ── OH   │
        |                                       |         │
   H ── C ── OH                            H ── C ───────┘
        |                                       |
     CH₂OH                                   CH₂OH

  Linear glucose                        Hemiacetal form
                                        of glucose
```

Figure 5.6 Cyclization of linear glucose to hemiacetal form

If the −OH group of the first carbon of the hemiacetal is on right side the sugar is called α-**form**; if the −OH group of the first carbon of the hemiacetal is on left side the sugar is called β-**form**. For example, α-D-glucose and β-D-glucose are shown in Figure 5.7.

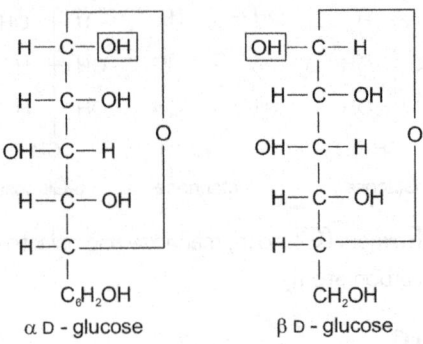

Figure 5.7   Structure of α-form and β-form of glucose

Ring Form

Ring form of aldoses resemble the structure of pyran ring (Figure 5.8). Hence, the ring form of aldoses are called pyranose form. Ring form of ketoses resemble the structure of furan ring (Figure 5.8):

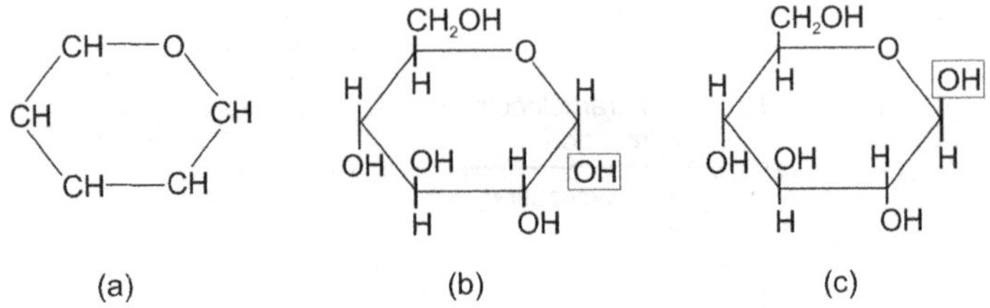

Figure 5.8   Structure of (a) Pyran ring (b) α-D-glucose and (c) β-D-glucose

Hence, the ring form of ketoses are called furanose form. If the −OH group of the first carbon of the ring is downward, i.e., below the plane, the sugar is called α-form; if the −OH is upward, i.e., above the plane, it is a β-sugar. For example, structure of α-D-glucose and β-D-glucose are shown in Figure 5.8. Structure of *α*-D-fructose and β-D-fructose are shown Figure 5.9.

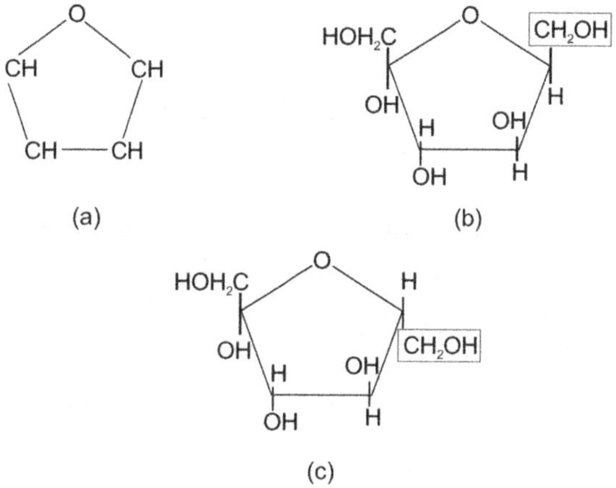

Figure 5.9 Structure of (a) Furan ring (b) α-D-fructose and (c) β-D-fructose

GLYCOSIDE LINKAGE

Sugar residues are linked to one another forming disaccharides, oligosaccharides and polysaccharides via glycoside linkage. For example, two glucose molecules are linked by glycoside bond forming a disaccharide, maltose. Formation of maltose through glycosidic linkage is shown in Figure 5.10.

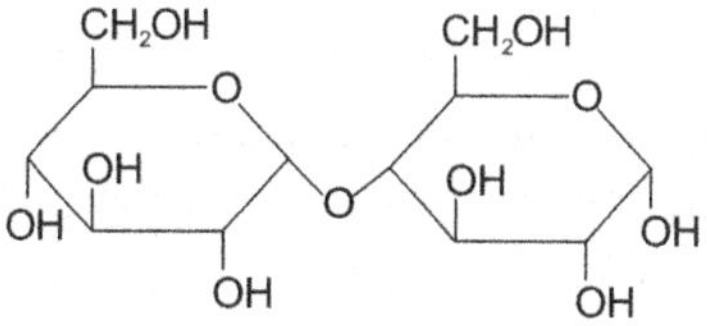

Figure 5.10 Glycoside bond between two glucose residues resulting in formation of maltose

Sucrose, a disaccharide is formed due to the formation of glycosidic linkage between glucose (aldose) and fructose (ketose) as shown in Figure 5.11.

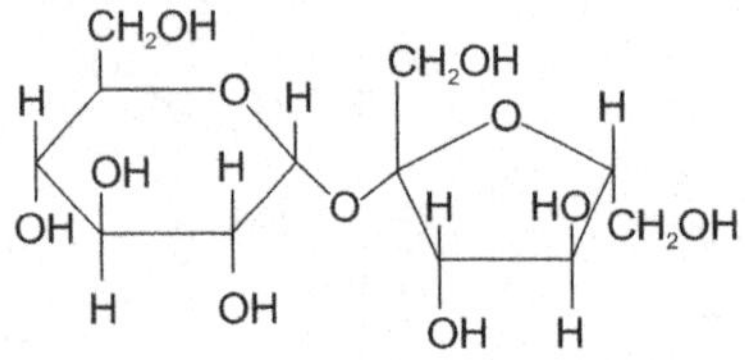

Figure 5. 1 1   Glycoside bond between one glucose residue and one fructose residue resulting in formation of sucrose

POLYSACCHARIDES

Polysaccharides are polymers of several sugar residues usually more than ten. There are two types of polysaccharides — homopolysaccharides   and heteropolysaccharides. Homopolysaccharides are composed of the same type of sugars, example, starch, glycogen and cellulose which are made up of only glucose units. Heteropolysaccharides are composed of more than one sugar or their derivatives.

Starch

Starch is a plant polysaccharide. It is a **storage** polysaccharide which occurs in the form of granules (Figure 5.12). It is structurally made up of two compounds — amylose and amylopectin (Figure 5.13). Amylose is made up of **linear** chain of α-D-glucose units linked by (α1 → 4) glycosidic linkage. Amylopectin is made up of branched chain of α-D-glucose units similar to amylose which are linked by (α1 → 6) glycosidic linkage at the branching point. Branching occurs at every 25–30 residues.

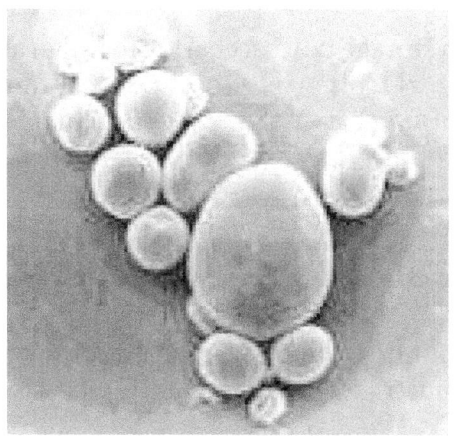

Figure 5.12 Stored starch granules

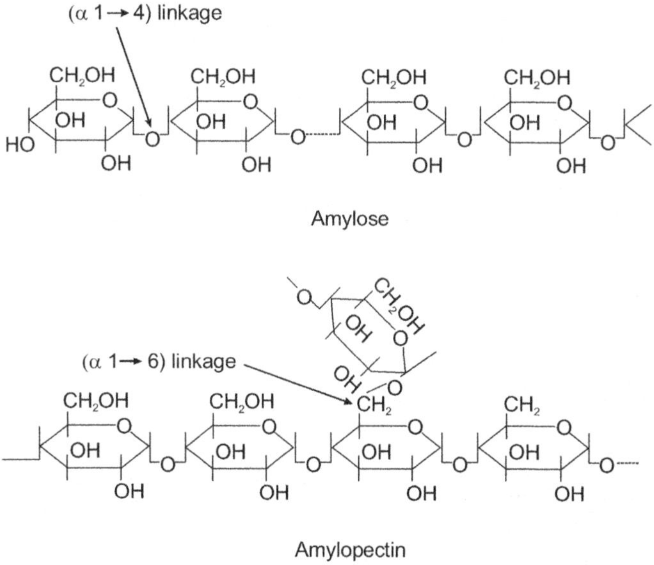

(α 1→4) linkage

Amylose

(α 1→6) linkage

Amylopectin

Figure 5.13 Structure of amylose and amylopectin in starch

Glycogen

Glycogen is a **storage** polysaccharide found in animals. The structure of glycogen is similar to the amylopectin component of starch except that the branching occurs more frequently, i.e., at every 8–12 glucose residues.

Cellulose

Cellulose is a **structural** polysaccharide made up of a linear chain of α-D-glucose units linked by (α1 → 4) glycosidic linkage. It is found in wood, paper and cotton. Structure of cellulose is illustrated in Figure 5.14.

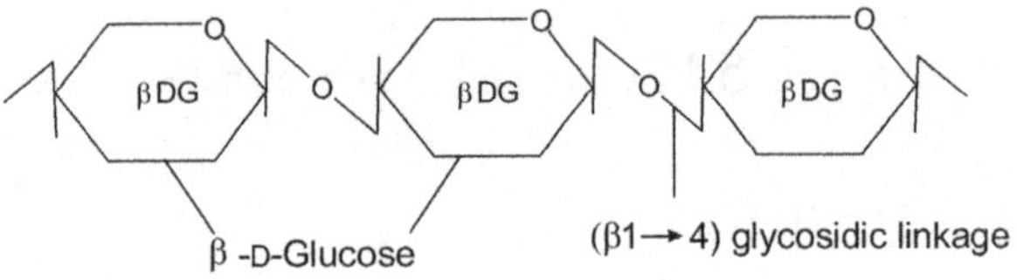

Figure 5.14   Structure of cellulose

Chitin

Chitin is a homopolysaccharide made up of N-acetylglucosamine units linked by (β1 → 4) glycosidic linkage. It is hard and tough and found in exoskeleton. It plays a **structural** role due to its high matrix strength and rigidity. Structure of chitin is shown in Figure 5.15.

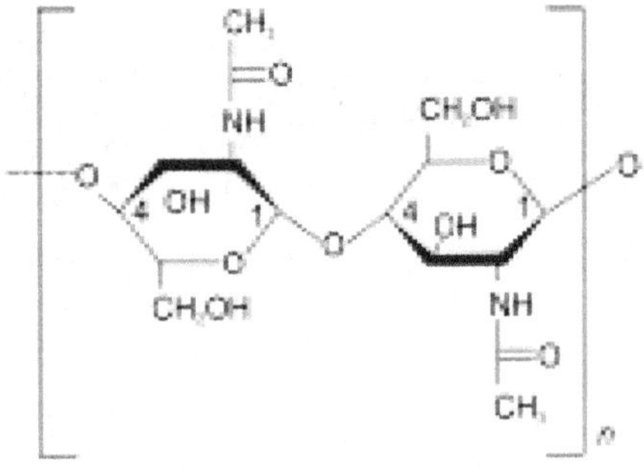

Figure 5.15   Structure of chitin

Inulin

Inulin is a polymer of fructose units. It is also called polyfructosan. It is found in plants typically in roots and rhizomes. It serves as a stored energy source.

CARBOHYDRATE DERIVATIVES

The following are some of the derivatives of carbohydrates and these have biological significance.

1. Sugar esters
2. Sugar acids
3. Sugar alcohol
4. Amino sugars
5. Deoxy sugars

Sugar Esters

The reaction of the —OH group of sugar with phosphoric acid gives sugar ester. This reaction is the initial step of sugar metabolism. Glucose 6-phosphate is an example for sugar ester (Figure 5.16).

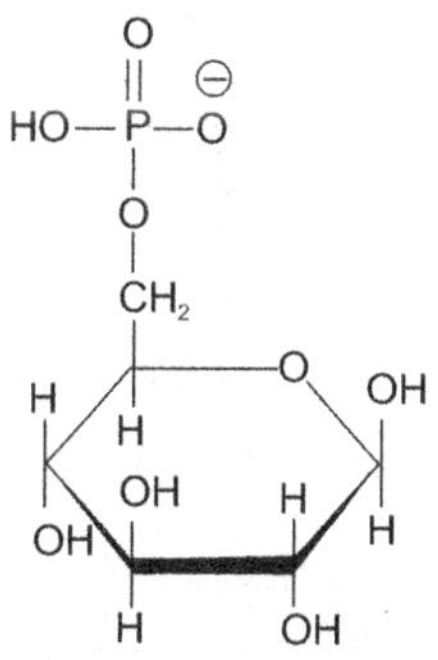

Figure 5.16   Structure of a sugar ester

Sugar Acids

Oxidation of aldehydic carbon or hydroxyl carbon of the sugar results in the formation of sugar acids. Examples are ascorbic acid which functions as vitamin and glucuronic acid which serves as structural component of proteoglycan as well as is involved in metabolism of bilirubin (Figure 5.17).

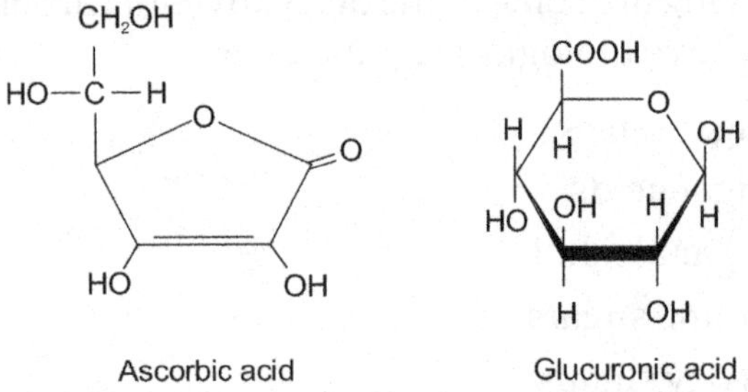

Figure 5.17   Structure of some sugar acids

Sugar Alcohol

Sugar alcohol is formed by the reduction of carbonyl carbon of aldoses and ketoses (Figure 5.18).

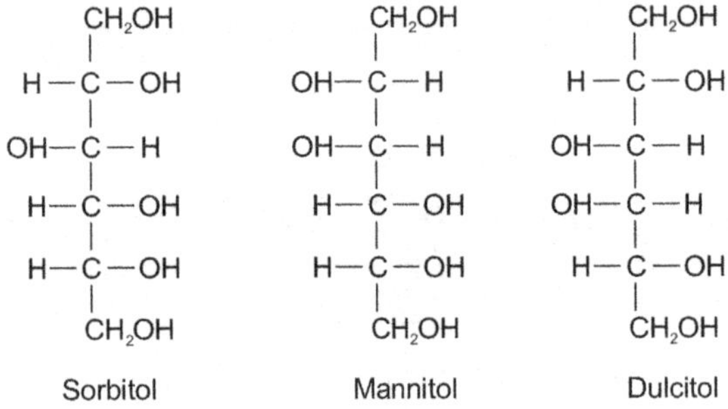

Figure 5.18   Structure of sugar alcohol

For example,

$$\text{D-Glucose} \rightarrow \text{D-Sorbitol}$$
$$\text{D-Mannose} \rightarrow \text{D-Mannitol}$$

D-Galactose → D-Dulcitol

D-Fructose → D-Mannitol + D-Sorbitol

Sugar alcohols are intermediates in minor pathways.

Amino Sugars

Amino sugars are formed when the hydroxyl group of a sugar is replaced by amino group or acetylamino group.

For example, glucose is substituted by amino group or acetylamino group to form D-glucosamine or *N*-acetyl-D-glucosamine respectively (Figure 5.19).

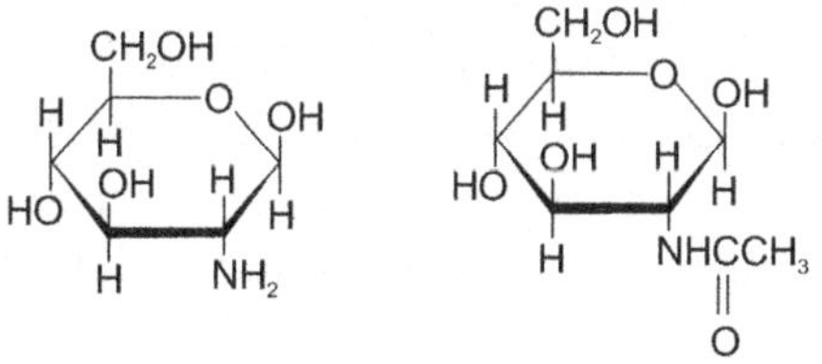

D-glucosamine   *N*-acetyl D-glucosamine

Figure 5.19   Structures of some amino sugars

Deoxy Sugars

Substitution of the hydroxyl group of a sugar with hydrogen atom yields deoxy sugars. For example, ribose can be converted into deoxyribose (Figure 5.20) which forms the structural component of DNA.

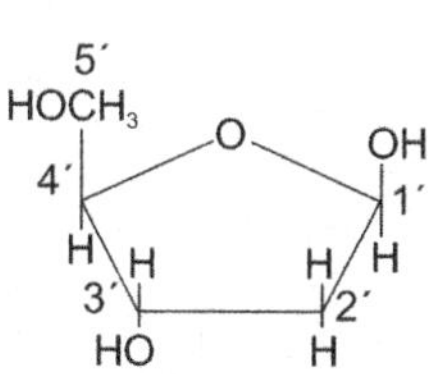

Deoxyribose

Figure 5.20   Structure of a deoxy sugar

HETEROPOLYSACCHARIDES

Heteropolysacchairdes are polymers of two different types of sugars or sugar derivatives. They are found in bacterial cell wall and extracellular matrix. Three types of heteropolysaccharides are known:

1. Peptidoglycan
2. Glycosaminoglycan
3. Proteoglycan

Peptidoglycan

Peptidoglycans are linear polymers of **N-acetyl muramic acid** (NAM) and **N-acetyl glucosamine** (NAG) linked by ($\beta1 \rightarrow 4$) linkage. NAM residues are attached to a tetrapeptide. Tetrapeptides are covalently cross-linked by a bridge of pentaglycine residues (Figure 5.21).

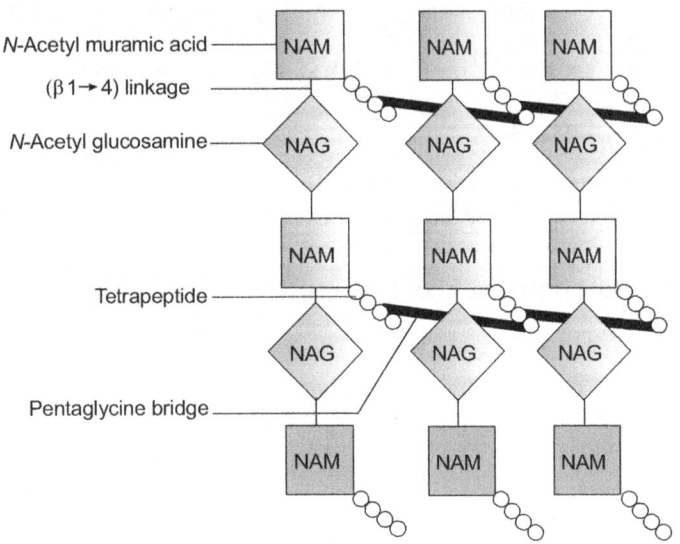

Figure 5.21   Structure of peptidoglycan

The ($\beta 1 \rightarrow 4$) linkage of peptidoglycan is susceptible to attack by an enzyme lysozyme which results in degradation of the polymer. This forms the basis for the protection of eyes from bacteria by the lysozyme present in tears, because, bacterial cell wall is made of peptidoglycan.

Glycosaminoglycan (GAG)

Glycosaminoglycan is a **linear** polymer of repeating disaccharide units in which one is always either **N-acetyl glucosamine** or N-acetyl galactosamine (amino sugar) and the other unit is usually **glucuronic acid**. In some cases the hydroxyl group of the amino sugar is **sulphated**. Due to the combination of carboxyl group of glucuronic acid and sulphates, GAG has high density of **negative** charge. GAG forms an interlocking porous **meshwork** in the extracellular matrix of connective tissues forming a **porous pathway** for the **diffusion** of nutrients and oxygen to individual cells. The different types of GAGs and their functions are shown in Table 5.2.

Proteoglycan

GAGs are attached to extracellular proteins to form aggregates called proteoglycans. The protein component of proteoglycan is called core protein. They are found in extracellular matrix.

GLYCOPROTEINS

Proteins covalently attached to carbohydrates are called glycoproteins. If the protein moiety is attached to the carbohydrate moiety through the hydroxyl group of serine or threonine residue, it is called O-linked glycoprotein. If the attachment is via amino group of asparagine residue, it is called N-linked glycoprotein.

Table 5.2   Glycosaminoglycans

GAG	Individual units	Function
Chondroitin 4-sulphate	*N*-acetyl galactosamine 4-sulphate D-glucuronic acid	Major structural component of cartilage and other tissues
Chondroitin 6-sulphate	*N*-acetyl galactosamine 6-sulphate D-glucuronic acid	Component of synovium and blood vessels
Dermatan sulphate	*N*-acetyl galactosamine 4-sulphate Iduronic acid	Structural component of skin
Heparin	*N*-sulpho galactosamine 4-sulphate D-glucuronic acid 2-sulphate	Anticoagulant
Hyaluronic acid	*N*-acetyl D-glucosamine D-glucuronic acid	Lubricant, shock absorber

Functions of Glycoproteins

They form the structural components of cell membrane and cell wall.

Mucus is a glycoprotein. It serves as lubricant.

Immunoglobulin is a glycoprotein which functions in immunity.

Human chorionic gonadotropin (hCG) is a glycoprotein hormone which plays a role in pregnancy.

Glycoproteins are involved in cell–cell attachment.

Examples of Glycoproteins

1. *Glycophorin* Glycophorin is the best characterized glycoprotein located in the red blood cell membrane. It has 16 oligosaccharide chains of which 15 are linked to protein via tyrosine/serine and 1 is linked via asparagine. The N-terminal end of the protein is present on the outer side of the cell membrane forming a hydrophilic domain or external domain. The C-terminal is located on the inner side of the membrane forming a hydrophobic domain or interior domain. The central part of the protein traverses through the membrane forming the transmembrane domain. The structure of glycophorin is shown in Figure 5.22.

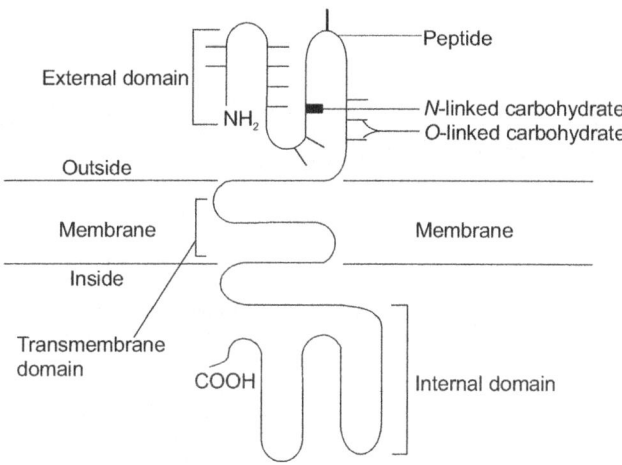

Figure 5.22 Glycophorin in red cell membrane

2. *Ceruloplasmin* Ceruloplasmin is a copper-transporting glycoprotein in the blood of humans and other vertebrates. It has oligosaccharide chains ending in sialic acid. Sialic acid determines the life of glycoproteins. If the sialic acid is removed, asialo-ceruloplasmin (meaning ceruloplasmin without sialic acid) is formed which binds to its specific receptor on the liver cells. Asialo-ceruloplasmin is degraded in the liver cells by lysosomes.

3. *Blood group antigens* Blood group antigens which determine the blood grouping of an individual are glycoproteins. Blood group antigens are located on the surface of red blood cells. There are three types of blood group antigens—A-antigen, B-antigen and O-antigen. All these three antigens have a common oligosaccharide core but differ in single sugar moiety attached to the common core.

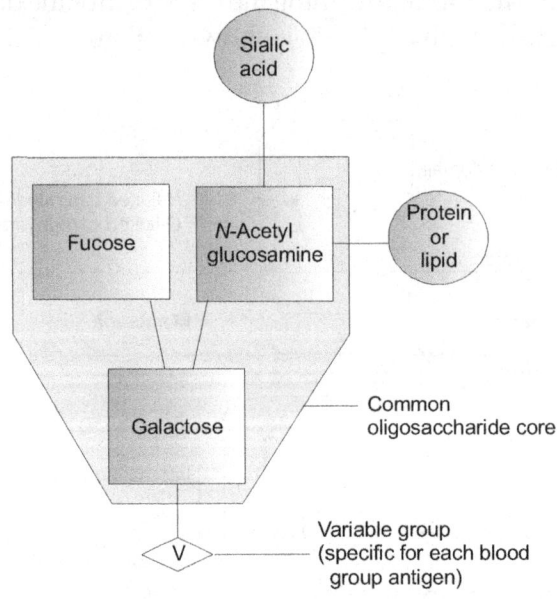

V = *N*-Acetyl galactosamine, for A group
V = Galactose, for B group
V is absent for O group

Figure 5.23 Structural components of blood group antigen (blood group glycoproteins)

Figure 5.23 illustrates the various components present in blood group antigen. The antigen has a core oligosaccharide made of galactose, *N*-acetyl glucosamine, fucose and sialic acid. *N*-acetyl glucosamine is linked to protein (forming glycoprotein) while galactose is linked to variable residue denoted as X. X is specific for each type of antigen and it determines the blood group of each individual:

X = *N*-acetylglucosamine, for A-antigen.

X = Galactose, for B-antigen

X is absent for O-antigen.

METABOLISM OF CARBOHYDRATES

Glucose is the major form of carbohydrate present in the biological system. Moreover, the end product of degradation of dietary starch, lactose and sucrose are glucose. Therefore, glucose metabolism assumes much significance. The various metabolic pathways of glucose are glycolysis, glycogenesis, glycogenolysis, gluconeogenesis, pentose phosphate pathway and Cori cycle of which glycolysis is most important as it forms energy.

VARIOUS METABOLIC PATHWAYS OF GLUCOSE
Glycolysis
Gluconeogenesis
Glycogenolysis
Glycogenesis
Pentose phosphate pathway
Cori Cycle

Glycolysis

Glycolysis is defined as the stepwise enzyme-catalysed oxidation of glucose with the net formation of ATP.

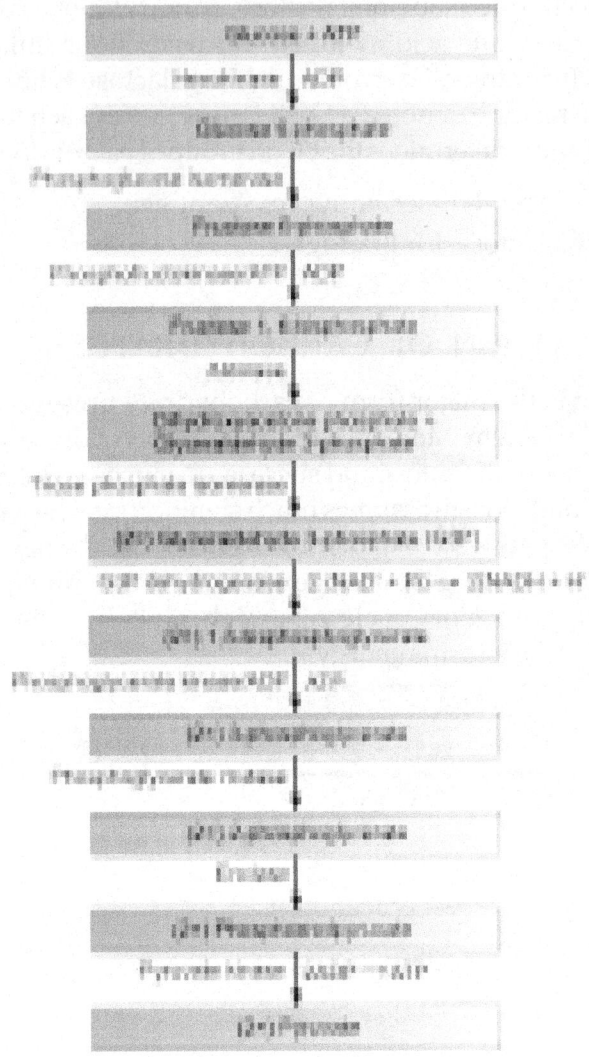

Figure 5.24 Steps involved in glycolysis (2× = 2 molecules)

Oxidation of glucose to lactate in anaerobic condition such as in muscle is called anaerobic glycolysis. Net ATP production in anaerobic glycolysis is 2 moles/mole of glucose.

Oxidation of glucose to pyruvate in aerobic condition such as in liver is called aerobic glycolysis. Net ATP production in aerobic glycolysis is 6 moles/mole of glucose.

The entire glycolytic pathway involves 10 steps. The first 4 steps form the first stage during which glucose is converted into triose phosphate. The last 6 steps form the second stage during which triose phosphate is converted to pyruvate. The steps involved in glycolysis are shown in Figure 5.24. Pyruvate formed in step 10 is converted into lactate by lactate dehydrogenase in anaerobic glycolysis.

Gluconeogenesis

Synthesis of glucose from non-carbohydrate precursors is called gluconeogenesis.

The overall reaction of gluconeogenesis is as follows:

$$2Pyruvate + 4ATP + 2GTP + 2NADH + 4H_2O \rightarrow$$

$$Glucose + 4ADP + 2GDP + 6Pi + 2NAD^+ + 2H^+$$

The non-carbohydrate precursors used in gluconeogenesis are:

- Lactate
- Pyruvate
- Glycerol
- Amino acids

Gluconeogenesis occurs more predominantly in liver and less predominantly in kidney and intestinal epithelial cells. This pathway is activated in starvation and during exercise.

Though the precursors are variable, the sequential metabolic pathway representing gluconeogenesis is just the reversal of glycolysis. But there are some enzymes which participate in gluconeogenesis but not in glycolysis; these enzymes are highlighted with asterisk symbol (Figure 5.25). Gluconeogenesis from different precursors are shown in Figure 5.26.

Figure 5.25   Gluconeogenesis pathway

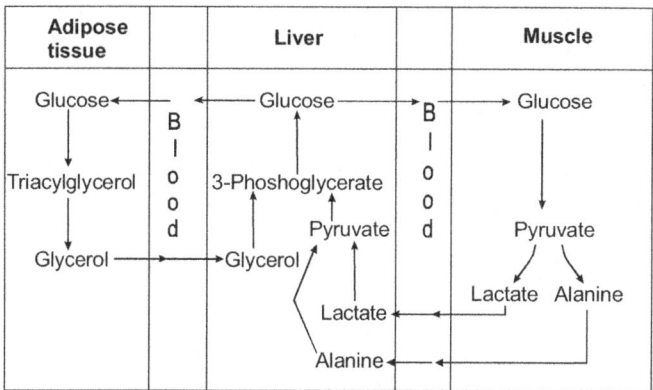

Figure 5.26   Gluconeogenesis from four different precursors

Gluconeogenesis from lactate/pyruvate—The Cori cycle
Lactate formed from anaerobic glycolysis in muscle is released into the blood and taken up by liver. Liver lactate dehydrogenase converts lactate to pyruvate.

$$\text{Lactate} + NAD^+ \rightarrow \text{Pyruvate} + NADH + H^+$$

Pyruvate is converted to 3-phosphoglycerate which is then converted to glucose and released into the blood. This glucose is used as energy source for muscle and other tissue. This process is called Cori cycle.

Gluconeogenesis from glycerol
During starvation, lipolysis takes place in the adipose tissue resulting in the conversion of triacylglycerol (TAG) into glycerol. Glycerol is released into the blood thereafter, and taken up by the liver where it is phosphorylated to 3-phosphoglycerate. phosphoglycerate is converted to glucose and released into blood.

Gluconeogenesis from amino acid
Alanine is converted to pyruvate which is converted to glucose. Alanine can also be converted to TCA cycle intermediates

which are metabolized to oxaloacetate an intermediate of gluconeogenesis (Figure 5.26 and 7.1).

Glycogenesis

Polymerization of glucose molecules to form long-chain polymer of glycogen is called glycogenesis.

$$\text{UDP-glucose} + \underset{\text{(Glycogen)}}{(\text{Glucose})_n} \rightarrow \text{UDP} + (\text{Glucose})_{n+1}$$

Glycogenesis occurs in fed state. Glycogen occurs as a storage polysaccharide in animals. It exist as granules in cytoplasm. Glycogen is stored in liver and muscle and whenever energy is needed, it is depolymerized to glucose.

Glycogenesis pathway starts with UDP glucose, the active form of glucose. UDP glucose is synthesized from glucose 1-phosphate. The pathway is shown in Figure 5.27 and schematically represented in Figure 5.28.

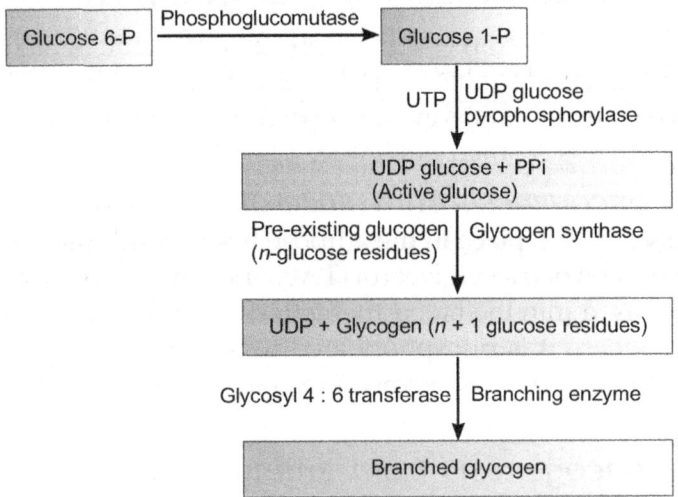

Figure 5.27   Glycogenesis pathway

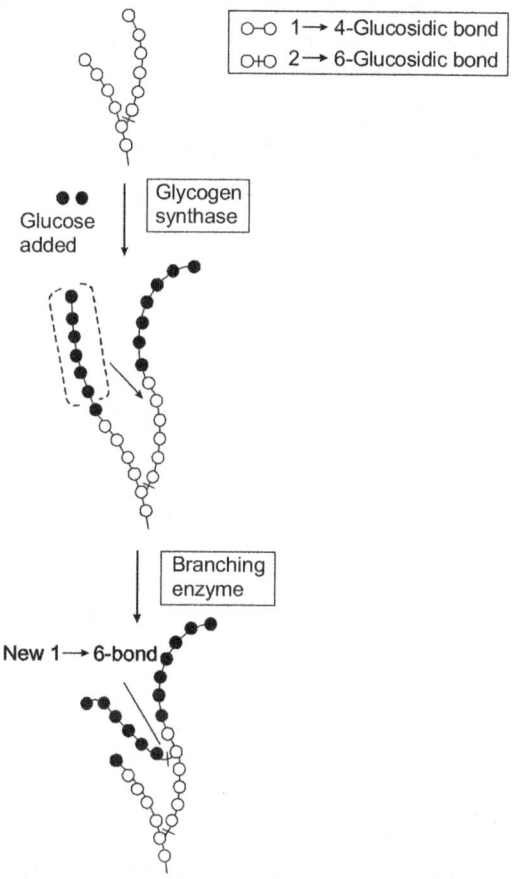

Figure 5.28 Schematic representation of glycogenesis pathway

Glycogen synthase adds UDP glucose to a pre-existing chain of glycogen with n-glucose residues to form a glycogen with $n + 1$ glucose residues. Glycogen is branched with the help of a branching enzyme called glycosyl 4 : 6 transferase. It introduces branching by **transferring** a terminal fragment of **6–7 residues** from a growing chain to a **6-position** farther back in a chain thus making a branch with an $(\alpha 1 \rightarrow 6)$ link creating two ends to add glucose.

Glycogen synthase, the crucial enzyme of this pathway, is activated by protein kinase which itself is activated by cyclic AMP (cAMP). cAMP is produced from ATP by adenylate cyclase. Adenylate cyclase is triggered by hormonal action. This activation cascade is shown below:

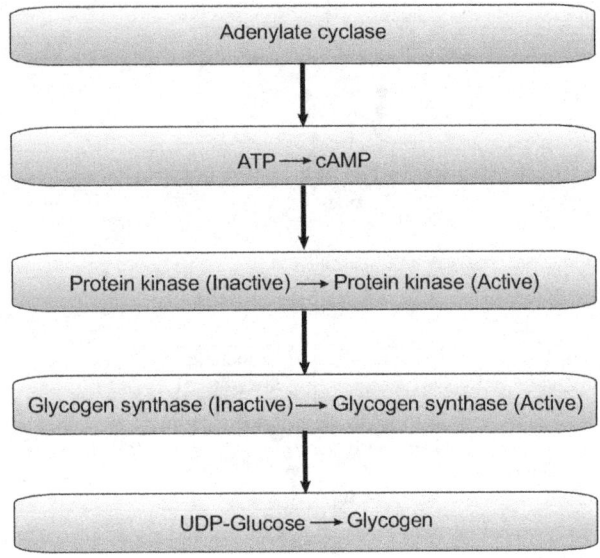

Glycogenolysis

Breakdown or depolymerization of glycogen into glucose units is called glycogenolysis.

Glycogenolysis involves phosphorolytic cleavage of $(\alpha 1 \to 4)$ glycosidic linkage by phosphorylase and debranching of glycogen by debranching enzymes—glucosyl transferase and amylo 6-glucosidase.

$$\text{Glycogen}(n\,\text{glucose residues}) + \text{Pi} \xrightarrow{\text{Glycogen phosphorylase}}$$
$$\text{Glycogen}(n-1\,\text{glucose residues}) + \text{Glucose 1-P}$$

Pentose Phosphate Pathway

Pentose phosphate pathway is a pathway for the interconversion of pentose and hexose. The reaction representing pentose phosphate pathway is:

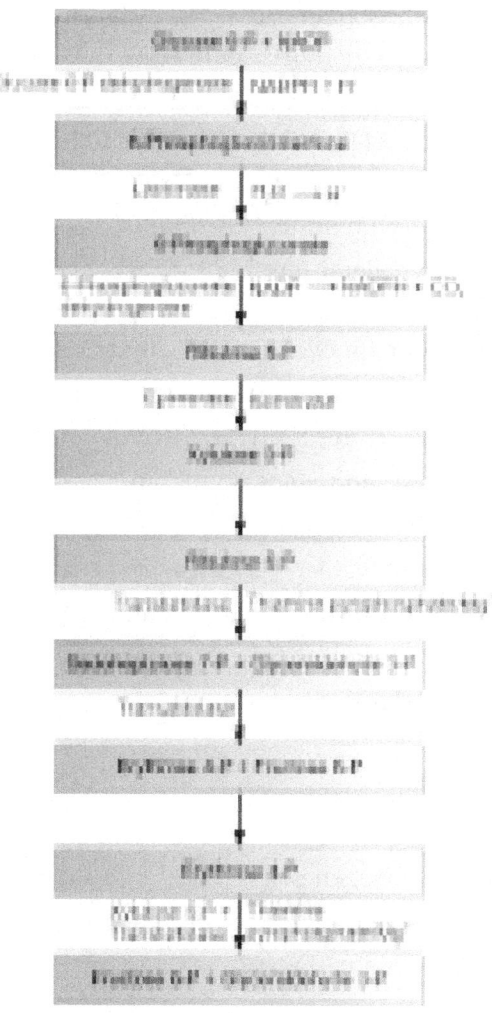

Figure 5.29 Pentose phosphate pathway

$$3 \text{ Glucose 6-P} + 6NADP^+$$

$$2\text{Fructose 6-P} + \text{Glyceraldehyde 3-P} + 6 \text{ NADH} + 6H + 3CO_2$$

This pathway is the source of NADH and ribose 5-P. It occurs in cytosol of liver, mammary gland, adipose tissue and erythrocytes. The pathway is shown in Figure 5.29.

REVIEW QUESTIONS

1. What are carbohydrates?
2. What is the general formula for carbohydrates? Give an example.
3. State any four functions of carbohydrates.
4. What is the structural role of carbohydrates?
5. Draw the structure of glyceraldehyde and glucose.
6. Define aldoses and ketoses with an example.
7. What is a chiral carbon?
8. Define enantiomers.
9. Draw the structure of furan ring and pyran ring.
10. Define epimers.
11. What is cellulose?
12. What is chitin?
13. What is inulin?
14. Define glycolysis.
15. Define gluconeogenesis.
16. What is glycogenesis and glycogenolysis?
17. What is pentose pathway?

CHOOSE THE CORRECT ANSWER

1. The most abundant sugar is
    (a) triose   (b) tetrose
    (c) pentose   (d) hexose

2. How many chiral carbons are present in glyceraldehydes?
    (a) one   (b) two
    (c) three   (d) four

3. Ring form of aldoses resemble the structure of
    (a) pyran ring   (b) furan ring
    (c) benzene ring   (d) all the above

4. Storage polysaccharide of plant is
    (a) glycogen   (b) cellulose
    (c) chitin   (d) starch

5. Glycophorin is present in
    (a) liver cells   (b) red cells
    (c) bone cells   (d) brain cells

Lipids are esters of alcohols and fatty acids, i.e., the two major constituents of lipids are alcohol and fatty acid. Fat, oil and wax are examples for lipids.

The alcohol component of lipids may be saturated (e.g. glycerol, cholesterol) or unsaturated (e.g. phytol). Fatty acids are carboxylic acid with a long saturated or unsaturated aliphatic chain. In general fatty acids have the molecular formula $CH-(CH_2)_n-COOH$. The fatty acid component of lipids may be of four types:

1. Saturated fatty acid (without double bond in its structure), e.g. palmitic acid and stearic acid.
2. Unsaturated fatty acid (with double bond in its structure), e.g. oleic acid, linoleic acid.
3. Hydroxy fatty acid, e.g. ricinoleic acid, cerebronic acid and dihydroxy stearic acid.

4. Cyclic fatty acid, e.g. hydnocarpic acid, chaulmoogric acid.

ESSENTIAL FATTY ACIDS

Some fatty acids are not present in the body but essential for normal cell functioning and should be taken through diet. These fatty acids are called essential fatty acids, e.g. linoleic acid, linolenic acid and arachidonic acid.

Isomerism due to unsaturated fatty acids Isomerism is a phenomenon by which a compound has two structural formulae but the same molecular formula. Unsaturated fatty acids exist in two different structural forms but with same molecular formula. They are *cis* form and *trans* form. For example, oleic acid is *cis* form and elaidic acid is *trans* form.

NOMENCLATURE OF FATTY ACIDS

The fatty acid chain has a CH_3 group at one end and a carboxyl (COOH) group at the other end. The number of carbon atoms in the fatty acid is numbered as C1, C2, etc. starting from the carboxylic end. Starting from C2, carbon atoms are given the Greek alphabetic notation. Thus, C2 is α-carbon, C3 is β-carbon, C4 is γ-carbon and so on.

$$\overset{}{CH_3} - \overset{}{CH_2} - \overset{}{CH_2} - \overset{\epsilon}{CH_2} - \overset{\delta}{CH_2} - \overset{\gamma}{CH_2} - \overset{\beta}{CH_2} - \overset{\alpha}{CH_2} - COOH$$
$$\;\;9\qquad 8\qquad 7\qquad 6\qquad 5\qquad 4\qquad 3\qquad 2\qquad\;\;\; 1$$

Each fatty acid is given a code number with three or four integers.

The first integer represents the total number of carbon atoms in the fatty acid chain.

The second integer represents the total number of double bonds in the chain.

The third integer represents the position of the double bond with respect to the serial number of the carbon atom bearing the double bond in the chain.

For example consider oleic acid with 18 carbon atoms.

$$(CH_3) - (CH_2)7 - CH_2 - \underset{9}{CH} = CH - (CH_2)7 - \underset{1}{COOH}$$

For oleic acid the code number is 18:1:9. It has 18 carbon atoms, 1 double bond in 9th carbon atom.

Consider linoleic acid. The code number is 18:2:9:12. It has 18 carbon atoms, 2 double bonds, one double bond in the 9th carbon atom and the other in the 12th carbon.

$$(CH_3) - (CH_2)4 - \underset{12}{CH} = CH - CH_2 - \underset{9}{CH} = CH - (CH_2)7 - \underset{1}{COOH}$$

FUNCTIONS OF LIPIDS

- Reserved source of energy.
- Provide calorific value.
- Structural components of cell membrane.
- Act as insulating component.
- Help in absorption and transport of fatty acids.
- Involved in biosynthesis of cholesterol, sex hormones and vitamin D.
- Carriers of fat-soluble vitamins.

PHYSICAL PROPERTIES OF LIPIDS

- Lipids are colourless, odourless and tasteless.
- Lipids are amphipatheic in nature, i.e., they have polar ($-COOH$) and nonpolar (CH_3) groups.

- Lipids are capable of absorbing odour.
- Lipids are insoluble in water and soluble in organic solvents.
- Lipids have high melting point which depends on degree of unsaturation. Melting point is inversely related to unsaturation.
- Lipids have specific gravity less than 1.
- They exhibit geometrical isomerism due to the presence of double bond in unsaturated fatty acid.
- Lipids have high insulating power.
- Lipids show emulsification property (formation of micelles).
- They function as precursors for synthesizing ketone bodies.

Emulsification is a process by which lipids combine with water or emulsifying agents like soaps, gums or proteins and form aggregates or droplets called micelles.

CHEMICAL PROPERTIES
Hydrolysis

Lipids are hydrolysed by lipase to form diacyl glycerol and monoacyl glycerol in two steps. Free fatty acid is liberated at each step.

$$\text{Lipids} \xrightarrow[\text{H}_2\text{O}]{\text{Lipase enzyme}} \text{Fatty acid + Diacyl glycerol} \xrightarrow[\text{H}_2\text{O}]{\text{Lipase enzyme}}$$

$$\text{Fatty acid + Monoacyl glylcerol}$$

Saponification or soap formation

The sodium salt of higher fatty acid gives hard soap, while the potassium salt of higher fatty acid gives soft soap.

$$\text{Fat + Alkali (NaOH/KOH)} \longrightarrow \text{Glycerol + Sodium salt of fatty acid (SOAP)}$$

Hydrogenation

Hydrogen is added to unsaturated fat in the presence of platinum/palladium catalyst. As a result saturated fat is formed.

Unsaturated fat $\xrightarrow[\text{H}_2]{\text{Platinum/Palladium}}$ Saturated fat (solid fat)
(vegetable fat) $(-CH_2-CH_2-)$
$(-CH=CH-)$

Halogenation

Halogen adds unsaturated fat forming saturated chloride derivative. This reaction is catalysed by acetic acid and methanol.

Unsaturated fat $\xrightarrow[\substack{\text{Acetic acid/} \\ \text{Methanol}}]{\text{Halogen}}$ Chloride derivative
$(-CH=CH-)$ $(-CHCl-CHCl-)$

Oxidation

Fatty acid is oxidized in two ways. In the presence of ozone aldehyde is formed. In the presence of $KMnO_4$ glycol is generated.

$-CH=CH- + \text{Ozone} \longrightarrow$ [cyclic structure] $\xrightarrow{H_2O}$ Aldehyde

$-CH=CH- \xrightarrow{\text{Mild } KMnO_4} \text{Glycol}$ [glycol structure]

Aldehyde formation

When fat is heated in the presence of $NaHSO_4$, acrolein (unsaturated aldehyde) is formed which is characterized by its pungent odour.

$$\text{Fat} \xrightarrow[\substack{\text{Dehydrating agent} \\ \text{NaHSO}_4}]{\text{Heat}} \begin{array}{l} \text{Unsaturated aldehyde} \\ \text{(Acrolein)} \end{array}$$

$$\downarrow$$

Pungent odour

Acid number   Acid number is the number of mg of KOH required to neutralize free fatty acid in 1 g of fat. This is used to quantify fatty acid in fat.

Saponification   number   (Sap   value/Sap number)   Saponification number is the number of mg of KOH required to saponify 1 g of fat. Sap value is inversely related to fatty acid chain length.

Iodine number   Iodine number is the number of grams of iodine absorbed by 100 g of fat. Iodine number is directly related to unsaturation.

Reichert–Meissl number (RM number/RM value)   RM number is the ml of 0.1N NaOH required to neutralize soluble fatty acid in 5 g fat.

CLASSIFICATION OF LIPIDS

Lipids are classified broadly into simple lipids, compound lipids and derived lipids. These lipids are sub-classified into different types. Classification and sub-classification of lipids are shown in Table 6.1.

Simple Lipids

Simple lipids are esters of fatty acids with the short chain alcohol glycerol. They are also called triacylglycerol or triglycerides and abbreviated as TAG or TG. The structure of glycerol and TAG or TG is shown in Figure 6.1.

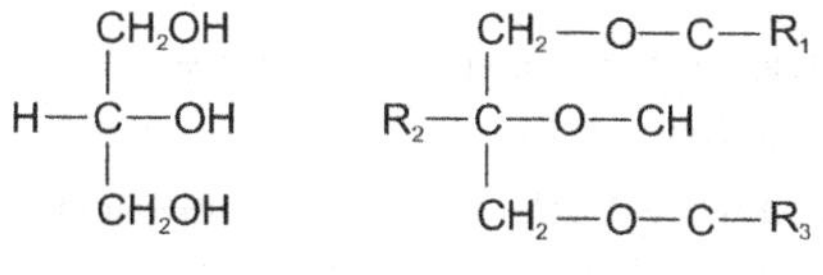

$$
\begin{array}{cc}
\text{CH}_2\text{OH} & \text{CH}_2\text{—O—C—R}_1 \\
| & | \\
\text{H—C—OH} & \text{R}_2\text{—C—O—CH} \\
| & | \\
\text{CH}_2\text{OH} & \text{CH}_2\text{—O—C—R}_3
\end{array}
$$

R = Variable alkyl chain

Glycerol   Triacylglycerol

Figure 6.1   Structure of glycerol and triacylglycerol (TAG)

Table 6.1   Classification of lipids

Simple lipids	Compound lipids	Derived lipids
Fats and oils (Triglycerides/ triacylglycerol)	**Phospholipids** Lecithin Cephalin Inositol Plasmalogen	**Sterol** Cholesterol
Waxes	**Sphingolipids** Sphingomyelin Sulphatid Cerebroside Ganglioside	**Bile acids** **Steroid hormones** Estrogen Androgen Mineralocorticoid Glucocorticoid Progesterone **Terpenes** Monoterpene Diterpene Triterpene **Carotenoids** Lycopene Carotene Xanthophyll

Fats and oils   Simple lipids with saturated fatty acids are called fats and they have solid consistency. They are also called animal lipids. Example, butter fat and tallow. Simple lipids with unsaturated fatty acids are called oils and they have liquid consistency. They are also called plant lipids. Example olive oil, castor oil, etc. Differences between Fat (animal lipid) and oil (plant lipid) are shown in Table 6.2.

Saturated fatty acid + Glycerol → Fat

Unsaturated fatty acid + Glycerol → Oil

Table 6.2   Differences between fat and oil (animal lipid and plant lipid)

Fat (animal lipid)	Oil (plant lipid)
Stored in liver and bone marrow	Stored in seeds and fruits
Solid nature	Liquid nature
Melting point is high	Melting point is low
Rich in saturated fatty acids	Rich in unsaturated fatty acids
Have low iodine value	Have high iodine value
Have high RM value	Have low RM value
Oxidative rancidity is more frequent	Oxidative rancidity is less frequent

Waxes   Waxes are hard esters of long-chain saturated fatty acid of 14–36 carbon atoms with long-chain monohydroxy alcohol of nearly 36 carbon atoms.

Long-chain saturated fatty acid + Long-chain monohydroxy alcohol → Waxes

Examples include lanolin wax from lamb wool, beeswax from honey comb and carnauba wax from Brazilian palm. Juniperic acid and palmitic acid are common long chain fatty acids found in wax. Cetyl alcohol, hexacosanol and myristyl alcohol are common long-chain alcohols found in wax. Waxes are fond in cutaneous glands, wool, fur and some leaves. They are highly resistant to oxidation.

Compound Lipids

Compound lipids are also called heterolipids. They are subdivided into phospholipids and sphingolipids.

Phospholipids Phospholipids are made up of glycerol, fatty acid and phosphoric acid which forms the phospholipid backbone (Figure 6.2a). They are also called phosphatidic acid.

Lecithin Phosphatidic acid with choline molecule attached to the phosphoric acid moiety is called lecithin. Lecithin is also called phosphatidyl choline. A simplified representation of lecithin is given in Figure 6.2b.

Cephaline Phosphatidic acid with serine or ethanolamine molecule attached to the phosphoric acid moiety is called cephaline. Cephaline is also called phosphatidyl serine or phosphatidyl ethanolamine. A simplified representation of cephaline is given in Figure 6.2c.

Inositide Phosphatidic acid with inositol molecule attached to the phosphoric acid moiety is called inositide. Inositide is also called phosphatidyl inositol. Simplified representation of inositide is given in Figure 6.2d.

Plasmalogen Lecithin or cephaline with an unsaturated ether is called plasmalogen. If the backbone is lecithin, it is called phosphatidyl serine/phosphatidyl ethanolamine. If the backbone is cephaline it is called phosphatidyl choline. Simplified representation of plasmalogen is given in Figure 6.2e.

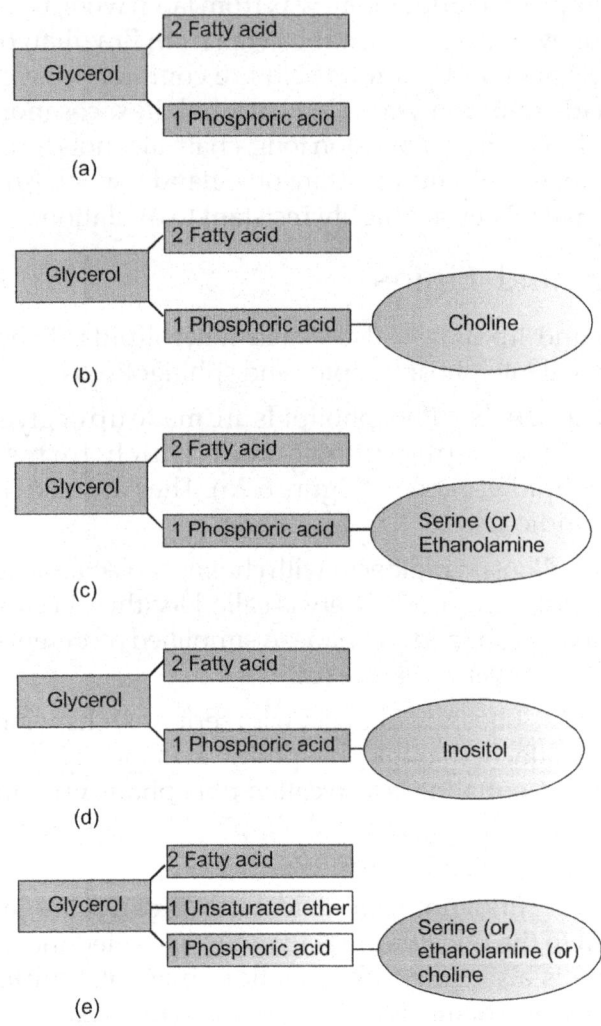

Figure 6.2　Simplified representation of phospholipids
(a) Phospholipid backbone (b) Lecithin (c) Cephaline
(d) Inositide (e) Plasmalogen

Lecithin (phosphatidyl choline) + Unsaturated ether →
Phosphatidyl choline (Plasmalogen)

Cephaline (phosphatidyl serine) + Unsaturated ether →
Phosphatidyl serine (Plasmalogen)

Cephaline (phosphatidyl ethanalomine) + Unsaturated
ether → Phosphatidyl ethanalomine (Plasmalogen)

Sphingolipids Sphingosine is made up of *N*-acetyl
sphingosine amide linked to fatty acid. This structure is
called sphingosine backbone (Figure 6.3a). Depending on
the additional groups attached to sphingosine backbone,
sphingolipids are sub-classified as follows:

Sphingomyelin Sphingosine backbone is linked to
phosphoric acid-choline to form sphingomyelin.
Sphingomyelin can be represented as in Figure 6.3b.

Sphingosine + Phosphoric acid-choline →

Sphingomyelin

Sulphatid Sphingosine backbone is linked to phosphoric
acid with sulphated galactose to form sphingomyelin.

Sphingosine + Phosphoric acid with

sulphated galactose → Sulphatid

A simplified representation of sulphatid is given in
(Figure 6.3c).

Cerebroside Sphingosine back bone is linked to glucose
or galactose to form sphingomyelin.

Sphingosine + Glucose / Galactose → Cerebroside

A simplified representation of cerebroside is given in
(Figure 6.3d).

Ganglioside   Sphingosine backbone is linked to glucose or galactose which in turn is attached to *N*-acetyl galactosamine + *N*-acetyl neuraminic acid to form sphingomyelin.

Sphingosine + Glucose / Galactose + *N*-acetyl

galactosamine + *N*-acetyl neuraminic acid → Ganglioside

A simplified representation of ganglioside is given in (Figure 6.3e).

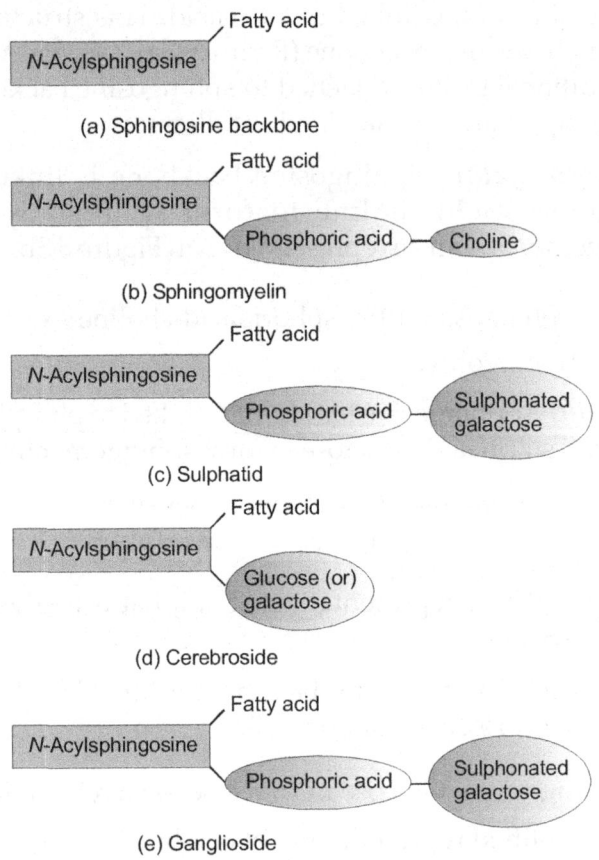

Figure 6.3   Simplified representation of sphingolipids

Derived Lipids

Steroids and sterols Steroids are lipids with four fused carbon rings. Fused ring structure is called **steroid nucleus** or **cyclopentanoperhydrophenanthrene** ring. Steroids with a hydroxyl group and an aliphatic chain of at least 8 carbons are called sterols. The structure of cyclopentanoperhydrophenanthrene.

Ring and cholesterol are shown in Figure 6.4.

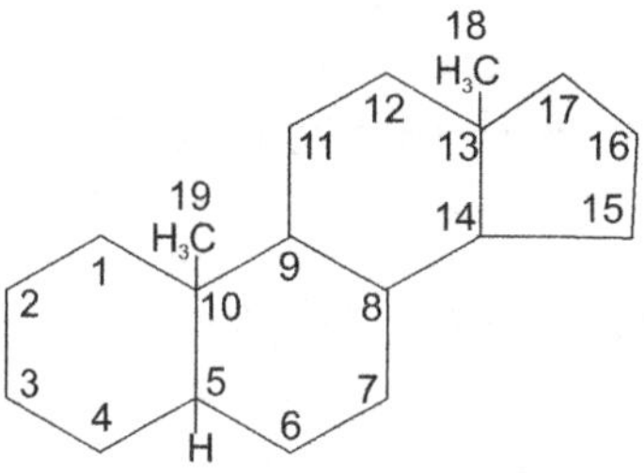

(a)

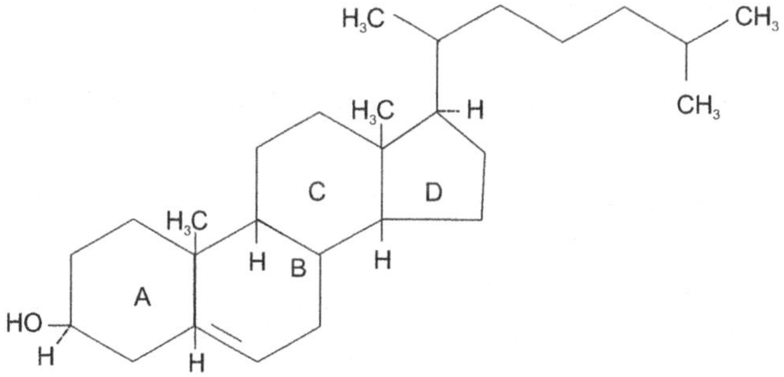

(b)

Figure 6.4 Structure of (a) Cyclopentanoperhydrophenanthrene (b) Cholesterol

Cholesterol Cholesterol is the major sterol in the human system. Structurally cholesterol is a steroid ring (cyclopentanoperhydrophenanthrene ring with OH group at 3rd carbon, double bond at 5th carbon and an aliphatic chain at 17th carbon (Figure 6.4a) Cholesterol is the precursor for bile acids, steroid hormones and vitamin D_3. Cholesterol is the structural component of cell membranes and plasma lipoproteins. Other sterols of interest are ergosterol present in yeast and ergot, lanosterol present in wool fat, and spinasterol present in spinach.

Bile acids   Bile acids are found in bile juice which is stored in gall bladder. They are derived from cholesterol. Thus, structurally, bile acids are steroids. Cholic acid, chenodeoxycholic acid, deoxycholic acid and lithocholic acids are predominant bile acid in humans. Cholic acid and chenodeoxycholic acid are called primary bile acids because they are directly derived from cholesterol, while deoxycholic acid and lithocholic acid are secondary bile acids because they are derived from primary bile acids. Bile acids are the major excretory pathways for cholesterol. Cholesterol is converted to bile acids and excreted in faeces.

Steroid   hormones   Estrogen,   androgen, mineralocorticoid, glucocorticoid and progesterone are collectively called steroid hormones. Steroid hormones are derivatives of cholesterol. The occurrence, and functions of these hormones are given in Table 6.3.

Terpenes   Terpenes are oxygenated derivatives of hydrocarbons. Building blocks of terpenes are called isoprene units, that may be represented in both ways as shown in Figure 6.5.

Figure 6.5   Structure of isoprene unit

Table 6.3 Steroid hormones

Name of the hormone	Occurrence	Function
Esterogen (Female sex hormone)	Follicle cells of ovary.	Regulates the activity of female reproductive system. Produces secondary sexual characters in females. Produces menopause symptoms. Prepares uterine mucosa for progestation hormone action. Increases skeletal muscle mass, growth and strength.
Androgen (Male sex hormone)	Leydig cells of testes.	Regulates the activity of male reproductive system. Produce secondary sexual characters in males.
Progesterone (Pregnancy hormone)	Corpus luteum. Placenta.	Helps in implantation of fertilized ovum. Aids uterine contraction. Induces maturation of mammary gland.
Mineralocorticoid	Cortical cells of kidney.	Regulates the metabolism of sodium and potassium. Regulates water distribution in body.
Glucocorticoid	Cortical cells of kidney.	Regulates the metabolism of carbohydrates, proteins and fats.

Based on the number of carbon atoms terpenes are called:

- Monoterpene $- C_{10}H_{16}$
- Sesquiterpene $- C_{15}H_{24}$

- Diterpene $-C_{20}H_{32}$
- Triterpene $-C_{30}H_{48}$
- Tetraterpene or carotenoids $-C_{40}$

Carotenoids Carotenoids are tetraterpenes with 40 carbon atoms and a high degree of unsaturation. They are also called lipochromes or chromolipids because they are coloured. Examples for carotenoids are carotene present in carrot, lycopene in tomato and xanthophyll present in corn and leaf.

METABOLISM OF FATTY ACIDS

The principal pathways involved in the biosynthesis and degradation of fatty acids are discussed here.

Biosynthesis of Fatty Acids

Fatty acid synthesis primarily takes place in the cytosol of liver cells.

Seven different enzymes catalysing fatty acid synthesis is clustered together to form a giant complex called fatty acid synthase complex. The seven enzymes of this multienzyme complex is listed below.

NAME OF THE ENZYME
Acetyl-CoA carboxylase(ACC)
Acyl-CoA-ACP transacylase (AAT)
Malonyl-CoA-ACP transacetylase (MAT)
β-ketoacyl-ACP synthase (BKAS)
β-ketoacyl-ACP reductase (BKAR)
β-hydroxyl acyl-ACP dehydratase (BHAD)
Enoyl-ACP reductase (EAR)

Glucose is converted to pyruvate via glycolysis. Pyruvate is converted to a two-carbon unit called acetyl-CoA. Acetyl-CoA is used as the precursor for fatty acid synthesis. Synthesis of fatty acids from acetyl-CoA is shown in Figure 6.6 and briefly explained here.

Acetyl-CoA has two functions. One molecule of acetyl-CoA forms malonyl-ACP (ACP is acyl carrier protein). Another molecule of acetyl-CoA forms acetyl-ACP.

Malonyl-ACP and acetyl-ACP combine to form butyryl-ACP through four steps.

Butyryl-ACP is the 4-carbon unit. It reacts with another malonyl group and the cycle continues. After completion of 7 cycles palmitic-ACP, a 16-carbon unit is formed.

Palmitic acid is released from the ACP and the fatty acid synthase complex by palmitoyl deacylase (thioesterase). The summary of the reaction catalysed by this complex is:

$$8\text{Acetyl-CoA} + 14\,\text{NADPH} + 14\text{H}^+ + 7\,\text{ATP} \rightarrow \text{Palmitate} +$$
$$8\text{CoA} + 14\,\text{NADP} + 7\,\text{ADP} + 7\,\text{Pi} + 7\text{H}_2\text{O}$$

Palmitic acid is elongated to 18-carbon stearic acid by reacting with malonyl-CoA.

Synthesis of Unsaturated Fatty Acid

Unsaturated fatty acids are synthesized from saturated fatty acids by introducing double bonds using the enzyme "mixed function oxygenase" in the presence of molecular oxygen and NADPH.

Fatty Acid Oxidation/ Fatty Acid Degradation

β-oxidation is the principal pathway for fatty acid degradation. It takes place in mitochondrial matrix.

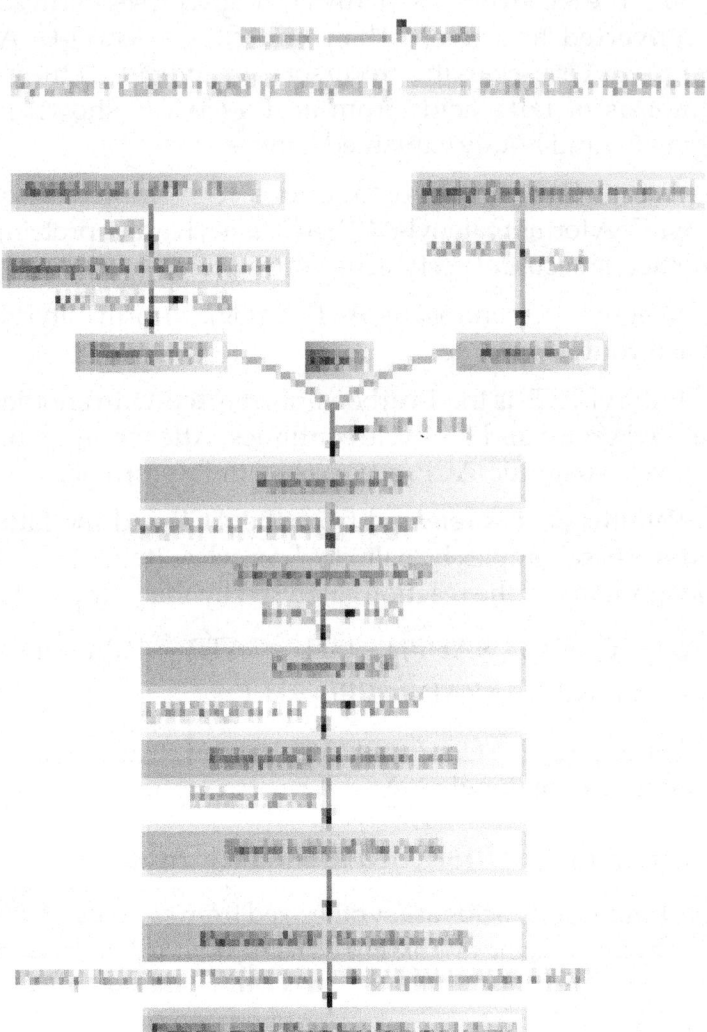

Figure 6.6   Synthesis of fatty acid (palmitic acid and stearic acid)

Fatty acids are activated to fatty acyl-CoA by the enzyme acyl-CoA synthetase in the cytosol.

$$\text{Fatty acid} + \text{ATP} + \text{CoASH} \xrightarrow{\text{Acyl-CoA synthetase}} \text{Acyl-CoA} + \text{PPi} + \text{AMP}$$

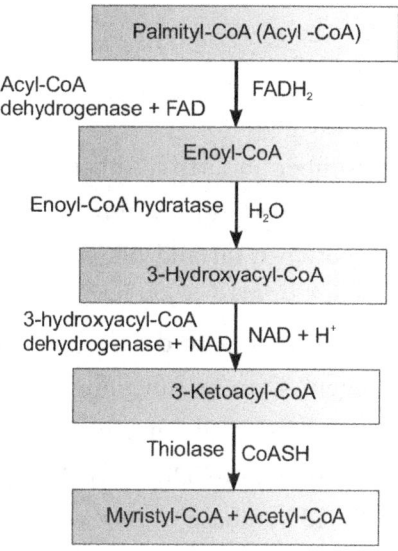

Figure 6.7   Oxidation of palmitoyl-CoA

Fatty acyl-CoA is transferred from cytosol to mitochondrial matrix, via the membrane-bound enzyme, carnitine palmitoyl transferase. Within the mitochondrial matrix, the fatty acyl-CoA is oxidized by 4 enzymatic steps. After the fourth step, a two-carbon unit is removed from the fatty acyl-CoA in the form of acetyl-CoA, forming a short-chain CoA. The short-chain compound again enters the degradation cycle. Thus in each cycle of oxidation two carbon units are removed. Figure 6.7 represents the β-oxidation pathway of palmitoyl-CoA.

REVIEW QUESTIONS

1. What are lipids? Give examples.
2. Write about isomerism in fatty acids.
3. Give examples for saturated, unsaturated, hydroxyl and cyclic fatty acids.
4. What is EFA?
5. List out the functions of lipids.
6. What is emulsification?
7. Define acid number and sap number.
8. Define iodine number and RM number.
9. Differentiate between fat and oil.
10. What are waxes?
11. Draw the structure of steroid ring and cholesterol.
12. What is estrogen? Give its function.
13. What is androgen? Give its function.
14. What is an isoprene unit? Draw its structure.
15. What is fatty acid synthase complex?
16. Give the function of Acyl-CoA synthetase and carnitine palmitoyl transferase.

CHOOSE THE CORRECT ANSWER

1. The molecular formula for fatty acid in lipids is
   (a) $CH-(CH_2)_n-COOH$
   (b) $CN-(CH_2)_n-COOH$
   (c) $CH-(CH_3)_n-COOH$
   (d) $CN-(CH_3)_n-COOH$

2. Lipids exhibit
   (a) optical isomerism   (b) geometrical isomerism
   (c) position isomerism   (d) structural isomerism
3. Sodium salt of higher fatty acid is called
   (a) soap   (b) cholesterol
   (c) sterol   (d) solid fat
4. Saponification value of fat is
   (a) inversely related to chain length
   (b) directly related to chain length
   (c) inversely related to unsaturation
   (d) directly related to unsaturation
5. The ring structure of cholesterol is called
   (a) cyclopentanoperhydrophenanthrene
   (b) cyclochloroperhydrophenanthrene
   (c) cyclopentanoperchlorophenanthrene
   (d) cyclopentanoperfluorophenanthrene

7 ELECTRON TRANSPORT CHAIN AND OXIDATIVE PHOSPHORYLATION

JOHN E. WALKER
Elucidated the enzymatic mechanism
underlying the synthesis of ATP.

Most eukaryotic cells and many bacteria are normally aerobic and they oxidize organic fuels completely to CO_2 and H_2O. Under these conditions pyruvate formed during glycolysis is oxidized to CO_2 and H_2O, via pyruvate dehydrogenase reaction and TCA cycle. During these two processes the energy released is conserved in the reduced electron carriers, NADH and $FADH_2$. These reduced cofactors are themselves oxidized giving protons and electrons. The electrons are transferred along a chain of electron-carrying molecules, known as respiratory chain or electron transport chain (ETC), to O_2 which is converted into water by reduction. Energy released during ETC is conserved in the form of ATP in the process called oxidative phosphorylation.

PYRUVATE DEHYDROGENASE REACTION

Pyruvate dehydrogenase (PDH) is an enzyme (multienzyme complex) which links glycolysis and the tricarboxylic acid (TCA) cycle. It is located in the mitochondrial matrix and catalyses the conversion of pyruvate, the end product of glycolysis, in the presence of NAD^+ to acetyl-CoA and NADH. Acetyl-CoA enters the TCA cycle and NADH enters electron transport chain to form the ultimate energy source ATP.

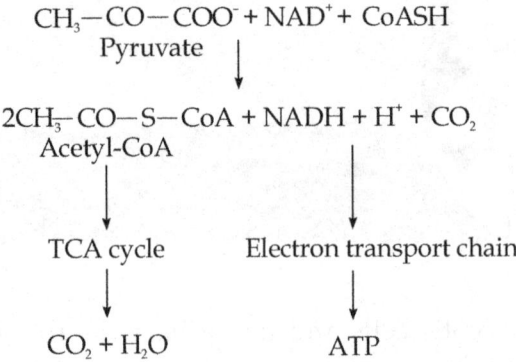

PDH is made up of 3 larger subunits, E1, E2, E3, possessing five types of enzyme activities, viz., pyruvate dehydrogenase, hydroxyl ethyl oxidase, 2-carbon unit transferase, dihydrolipoyl transacetylase and dihydrolipoyl dehydrogenase as shown in Table 7.1.

TCA CYCLE

TCA cycle (Tricarboxylic acid cycle) which is also called **citric acid cycle** or **respiratory cycle** or **Kreb's cycle** is a common pathway for the final oxidation of all metabolic fuels—carbohydrates, proteins, free fatty acids and ketone bodies. In short, all the fuels are converted ultimately into a two-carbon unit —**Acetyl-CoA**—which forms the substrate of the TCA cycle in the **mitochondrial matrix**.

Table 7.1   Pyruvate dehydrogenase multienzyme complex

Subunit	Enzyme activity	Cofactor
E1	Pyruvate dehydrogenase	Thymine pyrophosphate
E2	Hydroxyl ethyl oxidase	Lipoic acid
E2	2-Carbon unit transferase	Lipoic acid
E3	Dihydrolipoyl transacetylase	CoASH
E3	Dihydrolipoyl dehydrogenase	FAD/NAD

This pathway was discovered by Sir Hans H.A. Krebs. During the TCA cycle, the acetate group of acetyl-CoA is converted to CO_2 and H_2O. Sequential enzyme-catalysed reactions forming the TCA cycle are shown in Figure 7.1. Totally 12 ATPs are generated per cycle.

Carbohydrates (glucose), fatty acid (lipids) and some amino acids are converted to pyruvate. Pyruvate is converted to acetyl-CoA.

- Acetyl-CoA combines with oxaloacetate in the presence of citrate synthase forming citrate. This is a **condensation** reaction.

- Citrate is converted to isocitrate by aconitase by **isomerization** reaction.

- Isocitrate is converted to alpha ketoglutarate by **oxidative decarboxylation**. This reaction is catalysed by isocitrate dehydrogenase in the presence of **NAD** as a coenzyme.

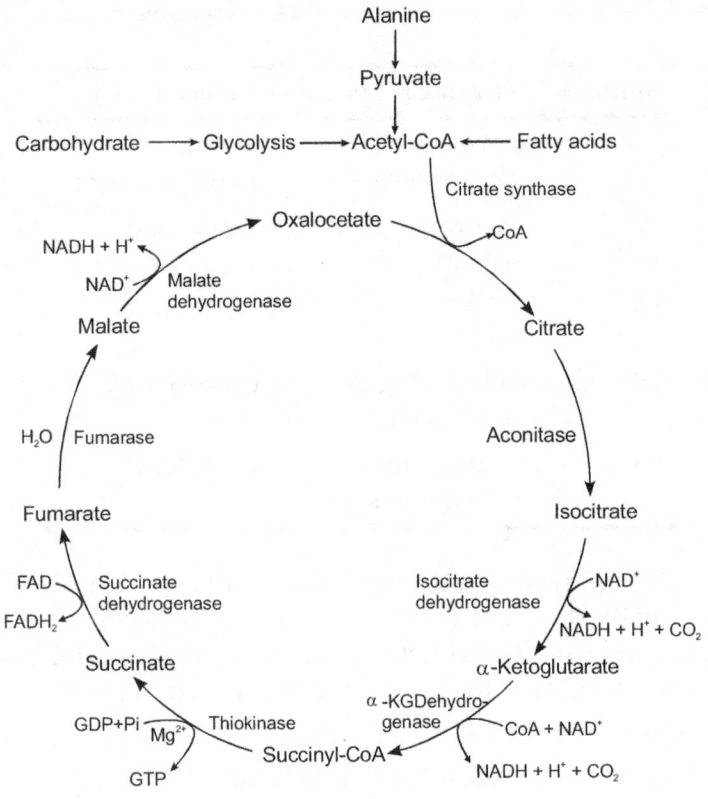

Figure 7.1   TCA cycle

- Succinyl-CoA is formed from alpha ketoglutarate in the presence of dehydrogenase, **Coenzyme A** and **NAD**. This is an oxidative decarboxylation reaction.

- Succinyl-CoA is converted to succinate by thiokinase. This reaction is called **substrate level phosphorylation** and takes place in the presence of GDP. Coenzyme A and **GTP** are formed in this reaction.

- Succinate is converted to fumarate by succinate dehydrogenase in the presence of **FAD**.

- Fumarase converts fumarate to malate in the presence of water.

- Malate is converted to oxaloacetate. This reaction is catalysed by malate dehydrogenase.

- Oxaloacetate again combines with the incoming acetyl-CoA and re-enters the cycle. Thus, TCA cycle generates **NADH, FADH$_2$** and **CO$_2$**. NADH and FADH$_2$ enter the respiratory chain and transfer their electrons to molecular oxygen and get themselves converted to NAD and FAD in order to re-enter the TCA cycle.

ELECTRON TRANSPORT CHAIN (ETC)

ETC is a common pathway in aerobic cells by which electrons derived from various substrates are transferred to oxygen. A scheme of mitochondrial electron transport chain is given in Figure 7.2. ETC is catalysed by a series of highly organized redox enzymes. It is composed of the following four complexes:

- **Complex I** — NADH-CoQ reductase (FMN, FeS)
- **Complex II** — Succinate-CoQ reductase (FAD, FeS, Cyt *b* 560)
- **Complex III** — CoQ–Cyt *c* reductase (Cyt *b*, FeS, Cyt *c*$_1$)
- **Complex IV** — Cyt *c* oxidase (Cu^{2+}, Cyt *a*, Cyt *a*$_3$)

All enzymes of ETC are integrated into the mitochondrial membrane. Functions of these enzymes are given in Table 7.2. NADH-CoQ reductase, passes electrons from NADH to coenzyme Q. Succinate-CoQ reductase, passes electrons from succinate to coenzyme Q. CoQ-Cytochrome *c* reductase transfers electrons to cytochrome *c* oxidase complex. Cytochrome *c* oxidase transfers electrons to oxygen, the terminal acceptor of electron in the electron transport chain forming water.

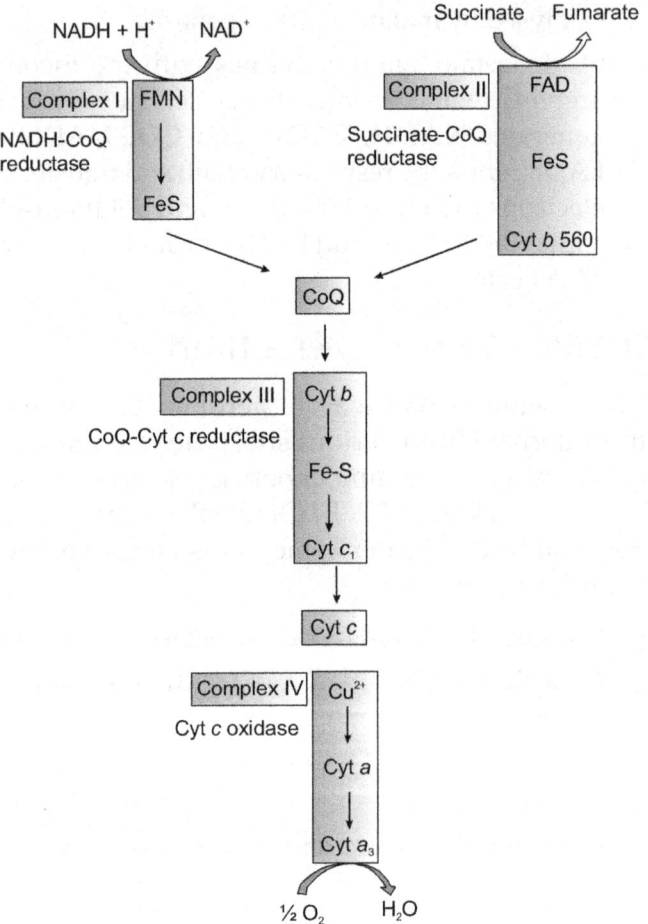

Figure 7.2   Electron transport chain

Table 7.2 Functions of enzymes in mitochondrial electron transport chain

Name of the complex	Function
NADH-CoQ reductase	Passes electrons from NADH to coenzyme Q
Succinate-CoQ reductase	Passes electrons from succinate to coenzyme Q
CoQ-Cytochrome *c* reductase	Transfers electrons from CoQ to cytochrome *c* oxidase complex
Cytochrome *c* oxidase	Transfers electrons from cytochrome *c* oxidase complex to oxygen, the terminal acceptor

OXIDATIVE PHOSPHORYLATION

The redox potential of electron carriers in ETC becomes more positive on going from NADH to molecular oxygen, i.e., the redox potential is in the following order:

$$-0.32V \rightarrow +0.04V \rightarrow +0.25V \rightarrow +0.82V$$

For each electron pair that passes through the ETC from NADH and $FADH_2$ 52 kcal of energy is released. This energy is coupled to the formation of ATP by phosphorylation of ADP. Coupling of oxidation of NADH + H^+ with the phosphorylation of ADP to form ATP is called oxidative phosphorylation. It is represented as in Figure 7.3. This process of oxidative phosphorylation is explained by chemiosmotic coupling hypothesis. According to this hypothesis, energy released during ETC is conserved by translocation of protons (H^+) from the mitochondrial matrix to the inter-membrane space. This translocation of proton creates an electrochemical gradient called as protomotive force (PMF). This PMF is harnessed to synthesize ATP by an enzyme **ATP synthase** (also called Complex V).

$$ADP + Pi + PMF \xrightarrow{\text{ATP synthase}} ATP$$

ATP synthase has two subunits:

F0 subunit which spans the membrane and forms a channel through which protons are channelized.

F1 subunit which is tightly bound to F0 subunit and bears the catalytic activity.

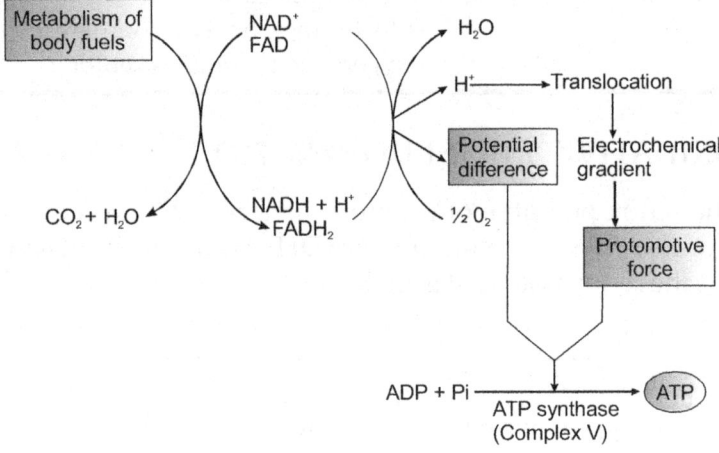

Figure 7.3 Simplified representation of oxidative phosphorylation

UNCOUPLERS

Uncouplers are a diverse group of compounds which inhibit the coupling between the electron transport and phosphorylation reactions and thus inhibit ATP synthesis without affecting the respiratory chain. Examples include Dicumarol, 2,4-dinitrophenol, chlorocarbonyl cyanide phenylhydrazone, oligomycin, carbon monoxide and rotenone.

These compounds permit leakage of protons, collapsing the proton gradient before it can be used for ATP synthesis. Some uncouplers are listed below.

UNCOUPLERS THAT INHIBIT ATP SYNTHESIS
Dicumarol
2,4-dinitrophenol
Chlorcarbomyl cyanide phenyl hydrazone
Oligomycin
Rotenone
Carbon monoxide

REVIEW QUESTIONS

1. What is the function of pyruvate dehydrogenase?
2. What is TCA cycle?
3. What is electron transport chain?
4. Write the function of the enzymes involved in electron transport chain.
5. What are uncouplers?

CHOOSE THE CORRECT ANSWER

1. Pyruvate dehydrogenase converts
   (a) pyruvate to acetate   (b) pyruvate to acetyl-CoA
   (c) acetate to pyruvate   (d) pyruvate to oxaloacetate

2. TCA cycle is also called
   (a) Kreb's cycle
   (b) respiratory cycle
   (c) citric acid cycle
   (d) all of the above

3. TCA cycle starts in
   (a) cytoplasm   (b) nucleus
   (c) mitochondria   (d) plasma membrane

4. Conversion of succinyl-CoA to succinate involves
   (a) dehydrogenation
   (b) oxidative phosphorylation
   (c) substrate level phosphorylation
   (d) reduction

5. In electron transport chain, the terminal electron acceptor is
   (a) hydrogen   (b) oxygen
   (c) nitrogen   (d) ATP

6. NADH-CoQ reductase is the
   (a) first enzyme of electron transport chain
   (b) first enzyme of TCA cycle
   (c) terminal enzyme of electron transport chain
   (d) terminal enzyme of TCA cycle

7. ATP synthase catalyses
   (a) oxidative phosphorylation
   (b) substrate-level phosphorylation
   (c) substrate-level deamination
   (d) oxidative deamination

Water is the universal solvent. It is present in the body as well as is supplied to the body by exogenous fluids and as water directly. The total body water comprises approximately 80% of the mass of most living cells. A normal adult will have 49 litres of water. Out of this, 35 litres are present in intracellular fluid (ICF) and 14 litres are present in extracellular fluid (ECF). Out of the 14 litres in ECF, 11 litres exist in interstitial tissues and 3 litres in blood plasma, cerebrospinal fluid, synovial fluid, and in the aqueous and vitreous humors. Water performs many functions: it transports, structures, stabilizes, lubricates, reacts and partitions.

STRUCTURE OF WATER

Atoms in water are arranged in a tetrahedral geometry. The H−O−H bond angle is 104.5°. Sharing of electrons between hydrogen and oxygen is unequal. Since oxygen is more electronegative, electrons are more often on the vicinity of the oxygen atom than hydrogen atom. Thus, the portion of the water molecule near the oxygen atom is slightly negative (δ-). Portion near the hydrogen atom is slightly positive (δ+) (Figure 8.1). Therefore, the water molecule is a dipole and it possess dipole moment.

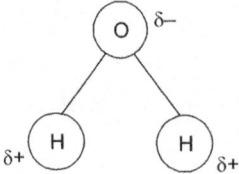

Figure 8.1   Structure of water molecule

The structure of the water molecule four valence orbitals of O point to corners of a tetrahedron. Two corners are orbitals with unshared pairs of electrons and weak negative charge. Two corners are occupied by H atoms which are in polar covalent bonds with O. Oxygen is so electronegative, that shared electrons spend more time around the O causing a weak positive charge near H's.

In liquid state, each water molecule forms hydrogen bonds with an average of 3.40 other water molecules. Continuous motion of water in liquid state indicates rapid formation and breakage of hydrogen bonds.

HYDROGEN BONDING IN WATER

The polar molecules of water are held together by hydrogen bonds. Hydrogen bonds in water are electrostatic interactions between hydrogen atom and the unshared electron pair of the electronegative atoms such as oxygen or nitrogen. Positively charged H of one molecule is attracted to the negatively charged O of another water molecule. Each water molecule can form a maximum of four hydrogen bonds with neighbouring water molecules. Hydrogen bonding orders water into a higher level of structural organization.

PROPERTIES OF WATER

Water molecules exhibit cohesive properties due to which they are held together by hydrogen bonds. Water has greater surface tension than most liquids. Water has high specific heat. (Specific heat is the amount of heat that must be absorbed or lost for one gram of a substance to change its temperature by one degree Celsius). Specific heat of water = One calorie per gram per degree Celsius $(1 \, cal/g/°C)$. As water cools from 4°C to freezing (0°C), it expands and becomes less dense than liquid water (ice floats). Water is a versatile solvent owing to the polarity of the water molecule. Dipole movements contribute to the solvent property of water. Water, being a dipole, interacts with positively and negatively charged ions forming a **water shell**.

Properties of water
Cohesiveness
High surface tension
Adhesion
High specific heat
High heat of vapourization
Freezing and expansion
Solvent property

FUNCTIONS OF WATER

Water molecules play an invaluable role in governing the structure, stability, dynamicity, and function of major biomolecules—proteins and nucleic acids. Indeed, these biomolecules lack activity in the absence of water. The living world should be thought of as an equal partnership between proteins, nucleic acids and water. Experimental information obtained from X-ray diffraction, neutron diffraction, nuclear magnetic resonance (NMR) and femtosecond spectroscopy measurements prove the active role of water in biomolecule structure, stability and dynamics.

Stabilization of Proteins by Water

Under normal pH, temperature and ionic strength conditions, proteins generally adopt a preferred three-dimensional (3D) fold. The hydrophobic effect and hydrogen bonds are also of prime importance in structuring proteins, and water is an important constituent in these contributions to protein structure and stability. The hydrophobic effect is generally considered to be the major driving force for the folding of globular proteins. It results in the burial of the hydrophobic amino acid side-chains in the core of the protein. Water tends to form ordered cages around non-polar groups (hydrophobic hydration) which leads to a decrease in the entropy of the system. These water molecules gain entropy when they are released after hydrophobic surfaces are put in contact with each other. This contributes in a very favourable way to the free energy of stabilization of the protein. Water is therefore fundamental in protein folding because of its role in defining hydrophobic attractions.

X-ray and NMR show that well-ordered water molecules are an integral part of all folded proteins, and water molecules have been found to be conserved among homologous proteins. Water molecules contribute to the structure and stability of proteins by bridging, via hydrogen-bonding different functional groups present in the protein. In certain cases buried water molecules contribute to the stabilization of protein conformations simply by filling pockets in the protein structure. Water not only induces protein folding and binding but also actively participates via water-mediated contacts. Water at the surface of biological macromolecules defines a layer, which has been termed "biological water". The hydration water in the proximity of the protein surface exhibits dynamical properties. Figure 8.2 schematically represents the way in which water is involved in protein structure and stability.

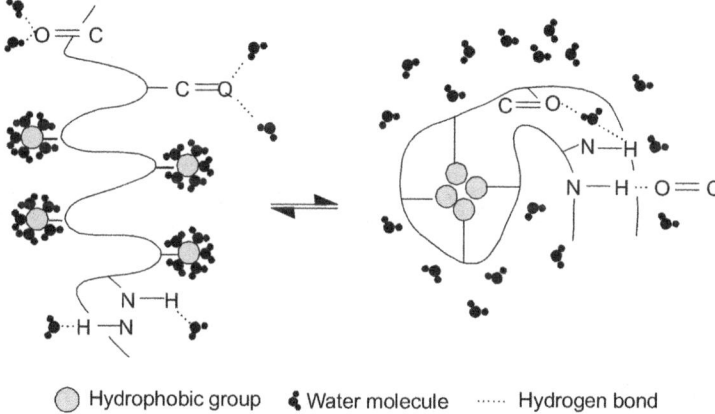

○ Hydrophobic group   ◖ Water molecule   ⋯⋯ Hydrogen bond

Figure 8.2   Role of water in maintaining protein structure and stability

Stabilization of Nucleic Acids by Water

Water plays a major role in stabilizing the structure of different forms of nucleic acids. Nucleic acids are highly charged polyanions. Along with hydrogen-bonding between base pairs and London dispersion forces between the stacked bases, water contributes to the stabilization of RNA and of DNA structures. The role of water in the stabilization of the 3D structure of nucleic acids is even more important than in proteins because of the presence of negatively charged phosphate groups. Phosphate–phosphate electrostatic repulsion is diminished in water by the high dielectric constant of water. The degree of hydration of nucleic acids also plays a key role in their conformation. For example high water activity favours the B form of DNA and reduced water activity leads to a transition from the B form to the C and A forms.

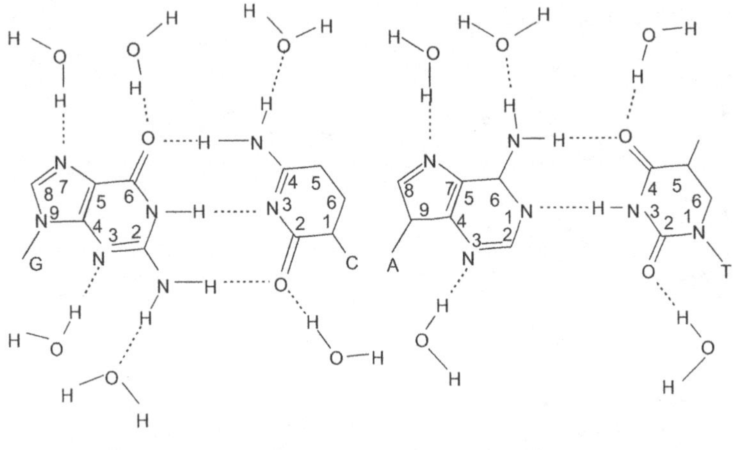

Figure 8.3   Water hydrogen-bonding to bases for stabilizing nucleic acid structure

Like in proteins, water is an integral part of nucleic acid structures. Water forms a hydration shell around DNA double helices. These water molecules are specifically bound to the phosphate groups and to the bases. X-ray studies show that there are six hydration sites per phosphate and that the positions and occupancies of these sites are dependent on the conformation and type of nucleotide. The paired bases are capable of hydrogen-bonding to water within the grooves (Figure 8.3). A spine of hydration occurs in the minor groove of B-DNA. This spine of hydration is the prime reason for the narrowing of the minor groove. RNA has a greater extent of hydration than DNA due to its extra oxygen atoms at the 2′ oxygen of ribose and unpaired base sites. Stable hydration patterns are observed, like in DNA around double-stranded regions.

Binding of Biomolecules

Binding processes are ubiquitous in biological systems, for example, binding of small molecules like drugs or binding of proteins to DNA. Binding processes generally occur in water. The fact that water molecules are abundantly observed experimentally at the interface between biomolecules suggests that water is indispensable for biomolecular recognition and self-assembly.

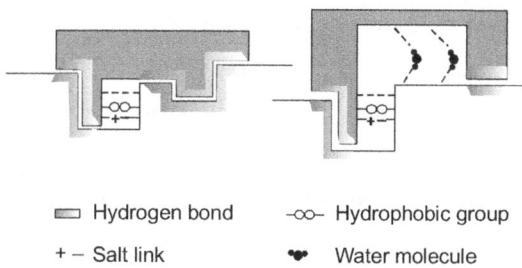

☐ Hydrogen bond   –∞– Hydrophobic group

\+ − Salt link   ❥ Water molecule

Figure 8.4   Water molecules promoting protein–ligand interactions

Water can act both as a hydrogen bond donor and acceptor, imposing few steric constraints on bond formation and can take part in multiple hydrogen bonds. Water plays a role in protein–ligand interactions allowing varied ligands to be accommodated in a given binding site or increasing affinity for specific ligands (Figure 8.4).

REVIEW QUESTIONS

1. Discuss the structure of water with reference to hydrogen bonding.
2. Discuss the properties of water.
3. Explain the role of water in stabilizing proteins.
4. How are nucleic acids stabilized by water molecules?
5. Give the biological significance of water.

CHOOSE THE CORRECT ANSWER

1. In water, atoms are arranged in
    (a) tetrahedral geometry
    (b) pentagonal geometry
    (c) trigonal geometry
    (d) cubical geometry
2. Water molecules are held together by
    (a) ionic bond   (b) hydrogen bond
    (c) disulphide bond   (d) hydrophobic bond
3. Water has
    (a) high specific heat and high cohesive property
    (b) high specific heat and low cohesive property
    (c) low specific heat and high cohesive property
    (d) low specific heat and low cohesive property
4. Phosphate–phosphate electrostatic repulsion of nucleic acid is diminished in water by
    (a) the high solubility of water.
    (b) the high dielectric constant of water

(c) the high cohesive property of water

(d) the high surface tension of water

5. Which of the following statements regarding water is not true?

(a) Water maintains the texture of the skin.

(b) Water reduces the risk of cystitis.

(c) Water functions as a shock absorber.

(d) Water functions as a catalyst.

9

VITAMINS

PAUL KARRER
Awarded Nobel Prize for his investigations on carotenoids, flavins, and vitamins A and B_2.

Vitamins are groups of organic molecules which are indispensable for the growth and maintenance of body and required in minute quantities. They should always be supplied via food.

Though vitamins are not synthesized by the body, they are indispensable for the growth and maintenance of the body. Hence vitamins are discussed as a part of biomolecules. A lack of vitamins in the diet causes deficiency diseases. Many of these deficiency diseases such as xerophthalmia, scurvy, beriberi and pellagra are very common all over the world particularly in the developing countries. The dietary vitamins requirement are necessary to prevent deficiency disorders. Requirement of vitamin is greater for growing children, and for pregnant and lactating women.

CLASSIFICATION OF VITAMINS

Vitamins are broadly classified into two categories based on their solubility.

1. Fat-soluble vitamins
2. Water-soluble vitamins

Vitamin A, vitamin D, vitamin E and vitamin K are fat-soluble vitamins. Vitamin B-complex (B_1, B_2, B_3, B_5, B_6, B_7, B_9, B_{12}) and vitamin C are water-soluble vitamins. Important details regarding fat-soluble vitamins and water-soluble vitamins are summarized in Tables 9.1 and 9.2 respectively.

FAT-SOLUBLE VITAMINS

Fat-soluble vitamins (A, D, E, and K) require a certain amount of fat in the diet for proper absorption, and excessive amounts of these vitamins can lead to toxicity.

VITAMIN A
(ANTIXEROPHTHALMIC VITAMIN)

Vitamin A is a fat-soluble vitamin stored in liver. It plays an important role in maintaining good eyesight and healthy skin.

Sources Following are the good sources of vitamin A namely, liver, carrots, mangoes, spinach, cantaloupe, dried apricots, milk, egg yolks and mozzarella cheese.

Structure Vitamin A occurs in many forms, such as retinol (alcohol), retinal (aldehyde), retinyl acetate or retinyl palmitate (esters), and provitamin A carotenoids (β-carotene, α-carotene, etc.).

Table 9.1   Fat-soluble vitamins

Vitamin	Common names	Chemical name	Major sources	Functions	Deficiency diseases
A	Antixerophthalmic vitamin	Retinol	Carrot, fish liver oil, butter, egg	Visual cycle Membrane integrity	Demyelinization, Nyctalopia Xerophthalmia Keratomalacia
D	Antirachitic vitamin Sunshine vitamin	Calciferol	Fish liver oil, egg.	Calcification of bones and teeth	Rickets Osteomalacia
E	Antisterilitic vitamin	Tocopherol	Wheat germ, egg	Antioxidant Nucleic acid metabolism	Sterility Muscular dystrophy Encephalomalacia
K	Antihaemorrhagic vitamin Coagulation vitamin	Phylloquinone	Spinach, cabbage	Prothrombin synthesis during blood clotting Oxidative phosphorylation	Haemorrhage Steatorrhoea

Table 9.2 Water-soluble vitamins

Vitamin	Common name	Chemical name	Major source	Function	Deficiency disease
B_1	Anti-beriberi vitamin	Thiamine or aneurine	Cereals	Coenzyme for decarboxylation reactions	Beriberi
B_2	Yellow enzyme	Riboflavin	Milk and green vegetables	Coenzyme for redox reactions	Glossitis Dermatitis
B_3	Pellagra-preventive factor	Niacin or nicotinamide	Yeast	Coenzyme for redox reactions	Pellagra Black tongue
B_5	Yeast factor	Pantothenic acid	Yeast, egg	Formation of acetyl-CoA coenzyme	Adrenal cortical insufficiency Depigmentation of hair
B_6	–	Pyridoxine	Cereals, grams, yeast, egg yolk	Growth factor Coenzyme in transamination reaction Carrier of amino acids	Convulsion Anaemia

B_7	Coenzyme-R Vitamin H Vitamin B_7	Biotin	Yeast, egg, peanuts	Coenzyme for carboxylase and fatty acid-synthesizing enzymes	Hair loss Oedema
B_9	Folacin	Folate	Yeast, meat	One-carbon-unit carrier Synthesis of purine, pyrimidine and amino acids	Megaloblastic anaemia
B_{12}	Antipernicious anaemia factor	Cyanocobalamin	Meat, fish	Nucleic acid metabolism	Pernicious anaemia
C	Antiscorbutic factor	Ascorbic acid	Citrus fruits, tomatoes	Reducing agent Collagen synthesis Synthesis of adrenal steroid hormones	Scurvy

Vitamin A is relatively unstable under normal storage conditions, particularly in harsh environments. The instability is mostly due to its chemical structure, which contains many double bonds susceptible to degradation. The structures of different forms of vitamin A are shown in Figure 9.1.

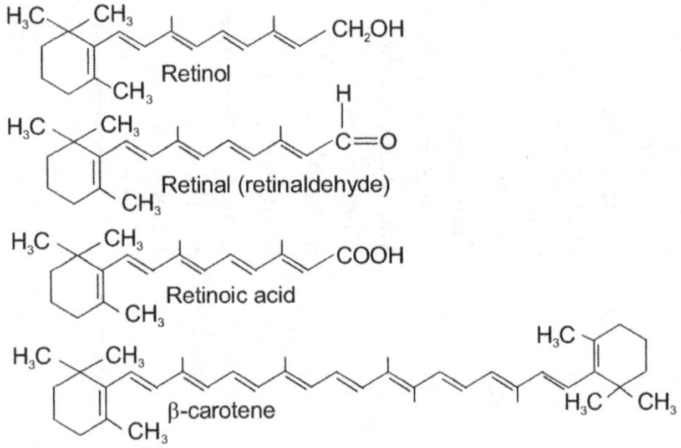

Figure 9.1   Structure of different forms of vitamin A

Functions   Vitamin A in the form of retinal is responsible for day vision (bright light) and night vision (dim light). Vitamin A functions in maintaining normal health of skin and mucosa. It also serves as an antioxidant.

Deficiency   An early sign of vitamin A deficiency is night-blindness. Severe vitamin A deficiency may result in permanent blindness. Vitamin A deficiency also causes abnormal function of many epithelial cells, manifested by such diverse conditions as dry, scaly skin, inadequate secretion from mucosal surfaces, infertility, decreased synthesis of thyroid hormones and elevated cerebrospinal fluid pressure due to inadequate absorption in meninges.

VITAMIN D (ANTIRACHITIC VITAMIN)

Vitamin D$_3$ is produced in skin when exposed to sunlight. It is also called antirachitic vitamin.

Sources Sources of vitamin D include milk, cod liver oil, mackerel, cheese, butter, margarine, fish.

Structure The two major forms of vitamin D are vitamin D$_2$ (ergocalciferol) and vitamin D$_3$ (cholecalciferol). The structures of these two forms are shown in Figure 9.2.

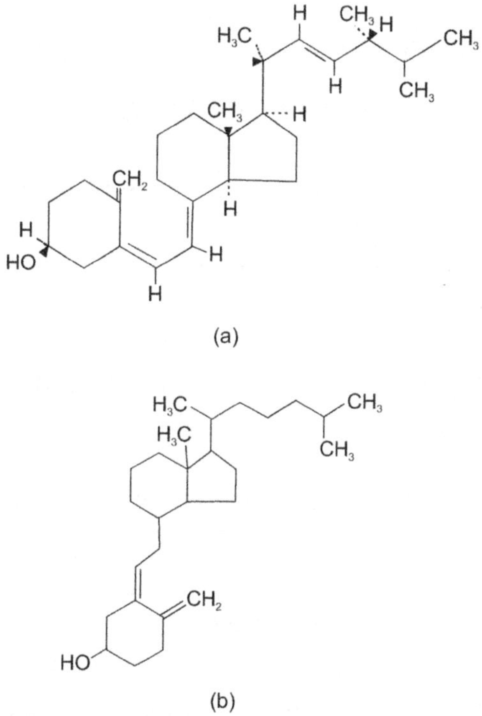

(a)

(b)

Figure 9.2 Structures of two forms of vitamin D (a) Ergocalciferol (b) Cholecalciferol

Functions Vitamin D promotes absorption of calcium and phosphorus in the intestine and reabsorption of calcium in the kidneys. Thus, it functions as a regulator of calcium and phosphorus levels in the blood. It also promotes mineralization of bone, bone growth and bone remodelling.

It can also promote soft-tissue calcification.

It is a stimulator of phagocytosis.

It exhibits anti-tumour activity and immunomodulatory activity.

Deficiency Deficiency of vitamin D results in

- Osteomalacia (softening and weakening of bones in adults).
- Rickets (softening and weakening of bones in children).
- Osteoporosis (porous and brittle bones easily susceptible to fracture).
- Tetany (Spasms of hand, feet, larynx, voice box due to low blood calcium (poor absorption of calcium).

VITAMIN E (ANTISTERILITIC VITAMIN)

Vitamin E refers to a family of related compounds called tocopherols. It is also called anti-sterility vitamin.

Sources Vitamin E is found abundantly in wheat germ, nuts, olives, green leafy vegetables, vegetable oil and sunflower seeds.

Structure Vitamin E has polar hydroxylated aromatic rings (chromanol rings) and non-polar isoprenoid side chains (Figure 9.3). Depending on the nature of side chains in the ring designated as R1, R2 and R3, different types of tocopherols exist. They are α-tocopherol, β-tocopherol, γ-tocopherol and δ-tocopherol. The molecule is lipophilic and resides almost exclusively in cell membranes.

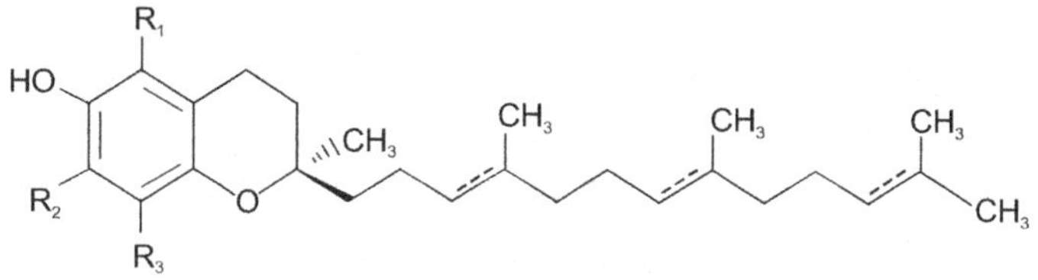

Compound	R_1	R_2	R_3	Phytyl chain
Alpha-tocopherol	CH_3	CH_3	CH_3	Saturated
Gamma-tocopherol	H	CH_3	CH_3	Saturated
Beta-tocopherol	CH_3	H	CH_3	Saturated
Delta-tocopherol	H	H	CH_3	Saturated
Alpha-tocotrienol	CH_3	CH_3	CH_3	Unsaturated
Gamma-tocotrienol	H	CH_3	CH_3	Unsaturated
Beta-tocotrienol	CH_3	H	CH_3	Saturated
Delta-tocotrienol	H	H	CH_3	Unsaturated

Figure 9.3   Structure of vitamin E (Tocopherol)

Functions   The primary function of vitamin E is maintenance of fertility. It helps proper functioning of sex organs.

Vitamin E is one of the important antioxidants, which intercepts free radicals and prevents cell damage. It particularly protects red blood cells.

It functions as a pain reliever, reducing menstrual pain, pre-menstrual syndrome, menopausal symptoms and muscle and joint pain.

Vitamin E also assists in the maintenance of vitamins A and C in the body.

Vitamin E supports healthy skin and assists in cell production. It protects against scar tissue and skin inflammation.

Antibody formation is enhanced by vitamin E.

Deficiency Low levels of vitamin E increase risk of declining physical function. Deficiency leads to sterility in males and a decrease in fat absorption.

VITAMIN K (COAGULATION VITAMIN OR ANTIHAEMORRHAGIC VITAMIN)

Sources Vitamin K is abundant in green vegetables like turnip green, cabbage, broccoli and cauliflower. Vegetables oils like soybean oil, cottonseed oil, olive oil, etc. are also good sources of vitamin K.

Structure Chemically, vitamin K molecules are 2-methyl 1,4-naphthoquinone derivatives. The structure of vitamin K is shown in Figure 9.4. It has a methylated naphthoquinone ring linked to a variable aliphatic side chain at its 3rd position. Depending on the side chain in its structure, vitamin K is divided into two important types — vitamin K_1 and vitamin K_2. Vitamin K_1 is **phylloquinone** which invariably contains four isoprenoid residues in its side chain, one of which is unsaturated (Figure 9.4a). Vitamin K_2 is called **menaquinone** and have side chains composed of a variable number of unsaturated isoprenoid residues; generally they are designated as MK-n, where n specifies the number of isoprenoids (Figure 9.4b).

Functions Vitamin K acts as coenzyme for glutamate carboxylase. It catalyses the carboxylation of γ-carbon of certain glutamate residues of proteins. The amino terminal glutamate residue of prothrombin (inactive) is carboxylated by vitamin K in the presence of ATP to form active thrombin which in turn coordinates with calcium ion for blood clotting.

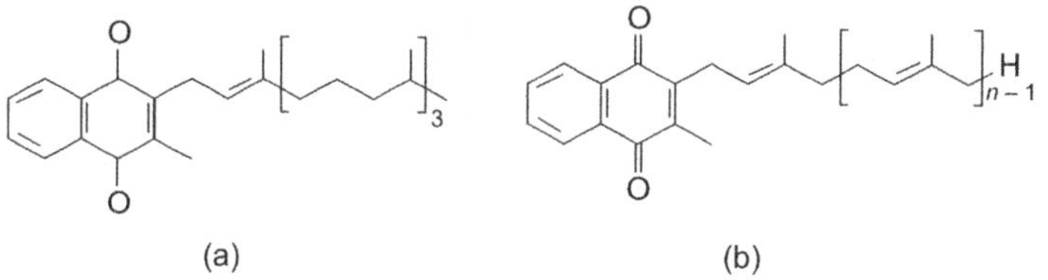

(a)   (b)

Figure 9.4   Structure of vitamin K (a) Vitamin K_1—phylloquinone (b) Vitamin K_2—menaquinone

Menaquinone intake is used to decrease the incidence of advanced prostate cancer.

Vitamin K is used as a prophylactic measure to prevent late-onset haemorrhagic disease.

Deficiency   Haemorrhage (blood loss) occurs due to vitamin K deficiency.

WATER-SOLUBLE VITAMINS

VITAMIN B_1 (THIAMINE)

Vitamin B_1, also called thiamine, is one of the B-complex vitamin. B-complex vitamins are necessary for healthy skin, hair, eyes, and liver. They also help the nervous system function properly.

Sources   The following foods are good sources of Vitamin B_1. Yeast, liver, pork, whole-grain cereals, rye and whole-wheat flour, wheat germ, navy beans and kidney beans, brown rice and lentils.

Structure   Thiamine is derived from a substituted pyrimidine and a thiazole which are coupled by a methylene bridge. Thiamine is rapidly converted to its active form, thiamine pyrophosphate, TPP, in the brain and liver by a specific enzyme, thiamine diphosphotransferase. The structures of thiamine and thiamine pyrophosphate are shown in Figure 9.5.

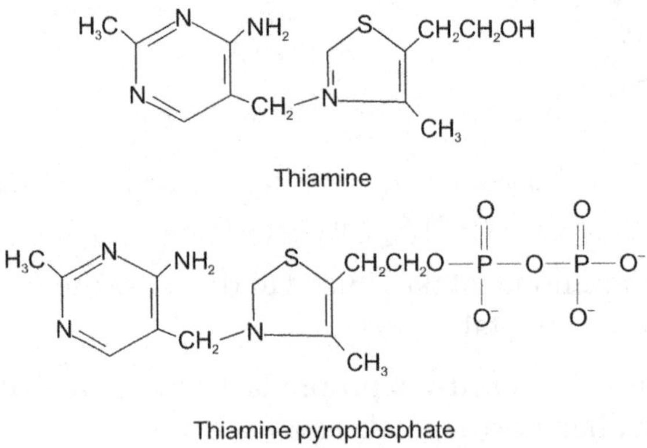

Thiamine

Thiamine pyrophosphate

Figure 9.5   Structures of thiamine and thiamine pyrophosphate

Function   Thiamine (vitamin B_1) helps the body cells convert carbohydrates into energy. It is also essential for the functioning of the heart, muscles, and nervous system.

Deficiency   Deficiency of vitamin B_1 leads to beriberi (nervous system ailment) and Wernicke–Korsakoff syndrome (a neurologic disorder characterized by acute encephalopathy followed by chronic impairment of short-term memory). Constipation, appetite suppression, nausea as well as mental depression, peripheral neuropathy and fatigue are symptoms of thiamine deficiency.

VITAMIN B_2 (RIBOFLAVIN)

Sources   Vitamin B_2 is found abundantly in the following – lean meats, eggs, legumes, nuts, green leafy vegetables, dairy products, and milk.

Structure Riboflavin is made up of **isoalloxazine** ring (the active part) attached to **ribitol** (Figure 9.6).

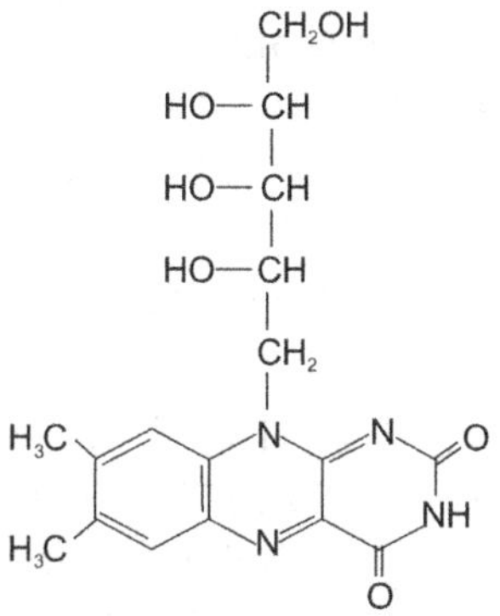

Figure 9.6 Structure of riboflavin

Functions Riboflavin is required for the metabolism of fats, ketone bodies, carbohydrates, and proteins. It functions as a coenzyme for redox reactions during metabolism. Riboflavin works with the other B vitamins. It is important for body growth and red blood cell production and helps in releasing energy from carbohydrates.

Deficiency Deficiency of riboflavin results in inflammation of the lining of mouth and tongue (glossitis), mouth ulcers, cracks at the corners of the mouth (angular cheilitis), sore throat, seborrhea (dandruff, flaking skin on scalp and face), trembling, sluggishness, and photophobia (excessive light sensitivity). Deficiency may also cause dry and scaling skin, fluid in the mucous membranes, and iron-deficiency anaemia. The eyes may also become bloodshot, itchy, watery and sensitive to bright light.

VITAMIN B$_3$ (NIACIN)

Niacin is vitamin B$_3$ which is also called pellagra-preventive factor.

Sources   Niacin is predominant in more animal products than plant products. Good sources of niacin include organ meats (kidney, liver), lean meats, pork, prawns, and even milk from a cow. Some of the other sources include seeds and almonds, rice bran, wheat products, beans, green vegetables, turnips, carrots, and celery. Yeast and bran can be good sources if the bran coating is not removed (as it is during the milling process).

Structure   Nicotinic acid and nicotinamide are two forms of niacin (Figure 9.7). Both nicotinic acid and nicotinamide can serve as dietary sources of vitamin B_3.

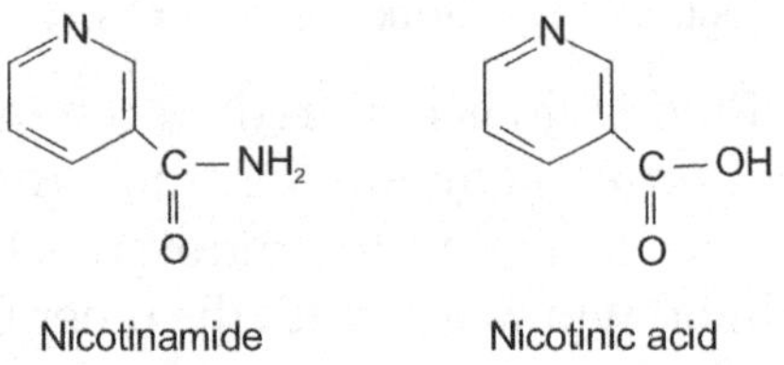

Nicotinamide   Nicotinic acid

Figure 9.7   Structure of nicotinamide and nicotinic acid (niacin)

Functions   Niacin is required for the synthesis of the two important coenzymes—nicotinamide adenine dinucleotide (NAD^+) and nicotinamide adenine dinucleotide phosphate ($NADP^+$). Both NAD^+ and $NADP^+$ function as cofactors for numerous dehydrogenases, e.g. lactate dehydrogenase and malate dehydrogenase and helps catalysing redox reactions.

Deficiency   Deficiency of niacin leads to a disease called pellagra which is characterized by overall feeling of tiredness, forgetfulness, nausea and/or vomiting, loss of appetite, skin outbursts or lesions, sores in the mouth, headache, anaemia, problems in the digestive system and trouble sleeping or relaxing.

VITAMIN B₅ (PANTOTHENIC ACID)

Pantothenic acid (pantothenate) is vitamin B₅.

Sources Whole grains, egg and yeast are good sources of pantothenic acid.

Structure Pantothenic acid is dihydroxydimethyl-butyric acid bonded to β-alanine. Structure of pantothenic acid is shown in Figure 9.8.

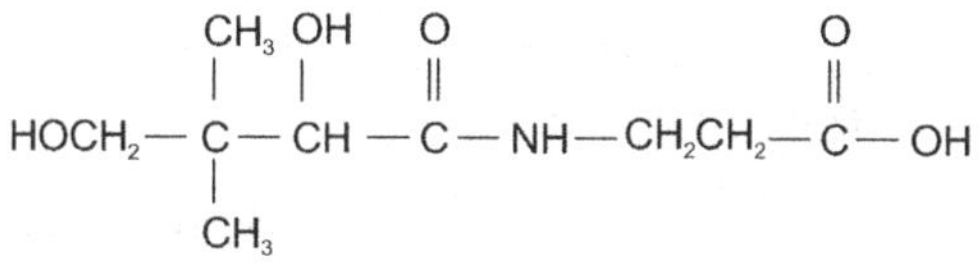

Figure 9.8 Structure of pantothenic acid

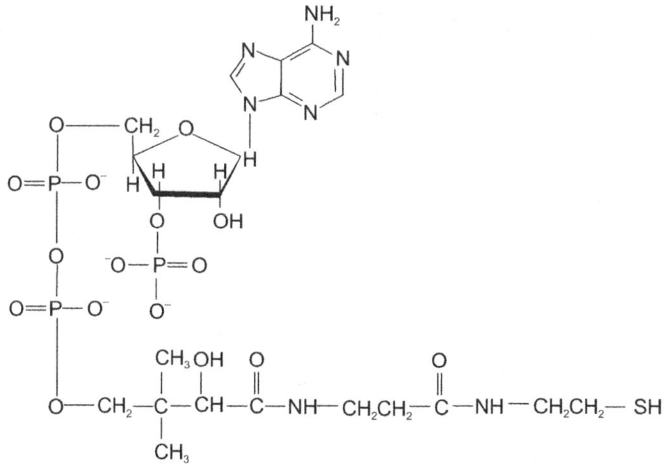

Figure 9.9 Structure of coenzyme A (derived from pantothenic acid)

Functions Pantothenate functions as a precursor for synthesis of coenzyme A (CoA) (Figure 9.9). It is also a part of the acyl carrier protein (ACP) domain of fatty acid synthase multienzyme complex (enzyme involved in fatty acid synthesis). Pantothenate is, therefore, required for the metabolism of carbohydrate via the TCA cycle and all fats and proteins.

Deficiency Deficiency of pantothenic acid is extremely rare due to its widespread distribution in whole grain cereals, legumes and meat.

VITAMIN B$_6$ (PYRIDOXINE)

Sources Vitamin B$_6$ is abundant in seeds, beans, bran and beef, egg, fish, and bananas.

Structure Pyridoxine has a pyridinium ring which possesses two alcoholic groups. Alcoholic groups are converted to aldehyde or amine groups to form pyridoxal or pyridoxamine (Figure 9.10). All the three compounds possess vitamin B$_6$ activity and are collectively called vitamin B$_6$.

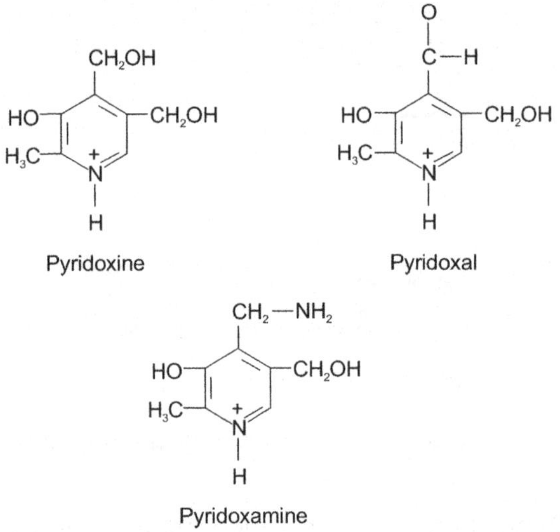

Figure 9.10   Structure of pyridoxine, pyridoxal and pyridoxamine

Functions Pyridoxal phosphate functions as a coenzyme for:

Transamination reactions.

Glycogen phosphorylase

The synthesis of the inhibitory neurotransmitter γ-aminobutyric acid (GABA).

Vitamin B_6 helps maintain normal nerve function and form red blood cells.

Vitamin B_6 has a major importance in regulating mood disorders and is the most implicated of all the vitamins in the cause and treatment of depression.

Deficiency Deficiency of vitamin B_6 is rare and it is usually associated with deficiency of vitamin B-complex. B6 deficiency syndrome includes nervousness, insomnia, skin eruptions, loss of muscular control, anaemia, mouth disorders, muscular weakness, dermatitis, arm and leg cramps, loss of hair, slow learning, and water retention.

VITAMIN B_7 (BIOTIN)

Biotin is vitamin B_7. It is also called coenzyme R or vitamin H.

Sources Rich sources of biotin are liver, kidney, yeast, milk products, egg yolks and nut meats.

Structure Structurally, biotin is made up of three moieties (Figure 9.11).

- Imidazole ring
- Thiophene ring
- Valeric acid moiety

Imidazole ring is *cis*-fused to a tetrahydrothiophene ring which is substituted at position 2′ by valeric acid.

Functions   Biotin functions as a coenzyme for several specific carboxylation and decarboxylation reactions.

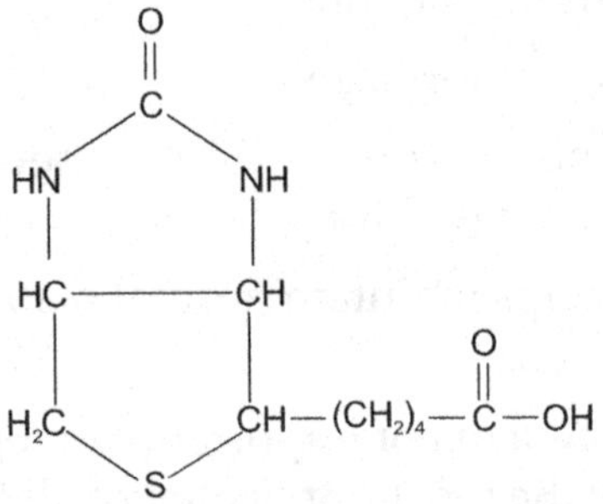

Figure 9.11   Structure of biotin

Biotin is also involved in the conversion of acetyl-CoA to malonyl coenzyme A in the formation of long chain fatty acids. Thus it is involved in fatty acid metabolism.

It is also involved in citrulline synthesis and may have effects on purine and pyrimidine synthesis.

Deficiency   Biotin deficiency results in skin disorders, muscle atrophy, lesions in the colon, loss of appetite, and spastic convulsions.

VITAMIN B$_9$ (FOLATE)

Folate (Folic acid) is called vitamin B$_9$ or folacin. Folate gets its name from the Latin word *folium* ("leaf").

Sources   Yeast, green vegetables, liver, kidney, glandular tissue, fish tissue, and fish viscera are good sources of folic acid.

Structure   Folic acid is a conjugated molecule (Figure 9.12) consisting of a pteridine ring structure linked to para-aminobenzoic acid (PABA) that forms pteroic acid. Folic acid itself is then generated through the conjugation of glutamic acid residues to pteroic acid.

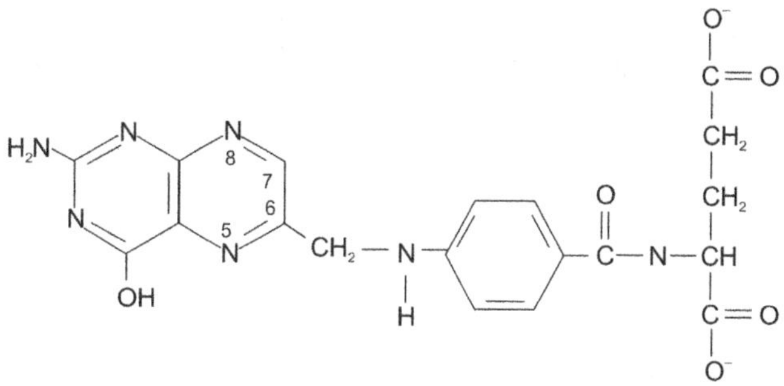

Figure 9.12   Structure of folic acid

Functions   Folic acid is required for normal blood cell formation.

It is the coenzyme in one-carbon transfer mechanisms such as serine and glycine interconversion, methionine-homocysteine synthesis, histidine synthesis, and pyrimidine.

It has a role in blood glucose regulation and improves cell membrane function.

Deficiency   Folate deficiency is characterized by poor growth, anorexia, general anaemia, lethargy, dark skin pigmentation and infarction of spleen.

VITAMIN B$_{12}$ (CYANOCOBALAMIN)

Cyanocobalamin is more commonly known as vitamin B$_{12}$.

Sources   Vitamin B$_{12}$ is found in liver, kidney, yogurt, dairy products, fish, clams, oysters, nonfat dry milk, salmon and sardines.

Structure   Cyanocobalamin bears a complex structure made up of **corrin** ring. The corrin has four **pyrrole** groups. A cobalt atom is present in the centre of the corrin ring and attached to the pyrole nitrogen. The corrin ring is attached to a cyano group (CN) on one side and a dimethylbenzimidazole ring on the other side. A simplified structure of cyanocobalamin is shown in Figure 9.13.

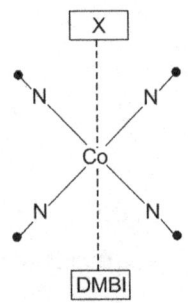

●—Pyrrole group
X—Cyano group or hydroxyl group or methyl group
DMBI—Dimethyl benzimidazole group
Co—Cobalt

Figure 9.13   Structure of cobalamin

Functions   Vitamin B_{12} is needed for building proteins in the body, red blood cells, and normal function of nervous tissue.

It functions as coenzyme for intramolecular rearrangement reaction and methyl group transfer reaction.

Deficiency   Pernicious anaemia and megaloblastic anaemia result from vitamin B_{12} deficiency.

Neurological complications also are associated with vitamin B_{12} deficiency and result from a progressive demyelination of nerve cells.

Deficiencies in B_{12} can also lead to elevations in the level of circulating homocysteine. Elevated levels of homocysteine are known to lead to cardiovascular dysfunction.

VITAMIN C (ASCORBIC ACID)

The term ascorbate is derived from the alpha privative "a" meaning "no" and "scorbuticus" meaning "scurvy" (the disease caused by deficiency of vitamin C). The molecular formula of ascorbic acid is $C_6H_8O_5$.

Sources   Following are the good sources of ascorbic acid. citrus fruits, green peppers, strawberries, tomatoes, broccoli, turnip greens, leafy greens, sweet and white potatoes, cantaloupe.

Structure   Ascorbic acid exists as D and L enantiomers. L-ascorbate is physiologically active and it is called vitamin C (D enantiomer is inactive). L-ascorbic acid is derived from glucose 6-phosphate during glucuronic acid pathway. The structure of ascorbic acid is shown in Figure 9.14.

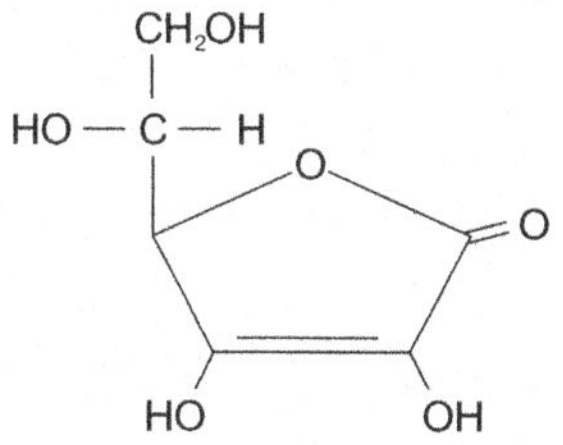

Figure 9.14   Structure of ascorbic acid

Functions   L-ascorbate functions as cofactor for certain hydroxylases, oxygenases and dehydrogenase.

In humans, vitamin C is a highly effective antioxidant, acting to lessen oxidative stress.

L-ascorbate functions as a coenzyme catalysing various biochemical reactions. Enzymes using L-ascorbate as cofactor are summarized in Table 9.3.

Table 9.3   Reactions catalysed by L-ascorbate

Enzyme	Process
Proline hydroxylase Lysine hydroxylase	Collagen synthesis
Trimethyllysine β-hydroxylase γ-butyro betaine hydroxylase	Carnitine synthesis
Dopamine- β-hydroxylase	Nor-epinephrine synthesis
4-OH-phenyl pyruvate dehydrogenase	Tyrosine metabolism
Peptidyl glycine α-monooxygenase	Amidation of peptide hormone

Deficiency   Deficiency of ascorbic acid causes a disease called scurvy. Signs of scurvy include tiredness, muscle weakness, joint and muscle aches, a rash on the legs, and bleeding gums, decreased wound healing and haemorrhaging, osteoporosis, and anaemia.

REVIEW QUESTIONS

1. How would you classify vitamins?
2. Write the functions of vitamin A.
3. Give a note on the functions of ascorbic acid.
4. Draw the structures of the following:
    i.   Pyridoxal
    ii. Niacin
    iii. Pantothenic acid

CHOOSE THE CORRECT ANSWER

1. Which of the following vitamins is involved in bone and teeth formation?

   (a) Vitamin A   (b) Vitamin D

   (c) Vitamin E   (d) Vitamin K

2. Vitamin E is called

   (a) antixerophthalmic vitamin

   (b) antirachitic vitamin

   (c) antisterilitic vitamin

   (d) antihaemorrhagic vitamin

3. Carboxylation of glutamate residues is catalysed by

   (a) vitamin B_1 (b) vitamin K

   (c) vitamin B_2 (d) vitamin E

4. Yellow enzyme is the other name for

   (a) vitamin B_1 (b) vitamin B_2

   (c) vitamin B_3 (d) vitamin B_5

5. Biotin is also called as

   (a) coenzyme P   (b) vitamin H

   (c) vitamin B_7 (d) all of the above

6. Which is the one-carbon-unit carrier?

   (a) Niacin   (b) Biotin

   (c) Folate   (d) Thiamine

7. Vitamin needed for collagen synthesis is

   (a) vitamin A   (b) vitamin B

   (c) vitamin C   (d) vitamin D

GLOSSARY

Activation energy Energy required for the formation of transition state during a biochemical reaction.

Active site Site on the enzyme in which the substrate binds.

Aldose Sugars with aldehyde group as the functional group.

Amino acid Building block of proteins and enzymes. They are made up of a central carbon atom linked to a hydrogen, an amino group, a carboxyl group and a variable group denoted as R.

Ampholytes Substances which have a dual nature, i.e., which can act both as an acid (proton donor) and as a base (proton acceptor).

Anabolism Biosynthesis of molecules in cells and part of metabolism.

Antioxidant A molecule that protects cells from oxidative damage caused by reactive oxygen and free radical molecules. Examples of dietary antioxidants are vitamins C, E, and K and diverse plant products such as lycopene, a nutraceutical found in tomatoes.

Bacteria Single-celled organisms and most prevalent forms of life on earth. Examples are *Escherichia coli, Salmonella typhimurium, Mycobacterium tuberculosis, Helicobacter pylori.*

Catabolism The part of metabolism responsible for degradation of nutrients and energy extraction for the benefit of ATP production.

Catalyst A substance that reduces the activation energy of a reaction, e.g. enzymes.

Catalytic power The ratio of the rate of catalysed reaction to the rate of uncatalysed reaction.

Cell The smallest independent part and structural and functional unit of an organism.

Cell cycle Complete sequence of steps which must be performed by a cell in order to replicate itself.

Cellulose Structural polysaccharide made up of linear chain of β-D-glucose units linked by (β1→4) glycosidic linkage.

Centrosome Dense region of cytoplasm that surrounds the nucleus.

Ceruloplasmin Copper-transporting glycoprotein in the blood of humans and other vertebrates.

Chargaff's rule As per this rule, DNA will have 1 : 1 ratio of purine and pyrimidine bases.

Chitin Homopolysaccharide made up of *N*-acetyl glucosamine units linked by (β1→4) glycosidic linkage.

Cloverleaf structure Tertiary structure of RNA with four arms and one loop.

Contractile proteins Proteins which perform mechanical work like muscle contraction, e.g. actin and myosin.

Cytoskeleton A fibrous network made of proteins that contributes to the structure and internal organization of eukaryotic cells. The cytoskeleton is found in the cytoplasm of cells and has three major fibre types: microfilaments made of actin protein, intermediate filaments made of various kinds of proteins (e.g. keratin), and microtubules made of tubulin.

Defence proteins Proteins which provide defence by invading bacteria and other foreign organisms, e.g. immunoglobulin.

Dimorphism Capacity of pathogenic (disease-causing) fungi to exist in two forms, yeast form (Y) and mould (M) form.

DNA Deoxyribonucleic acid. It makes up the genetic component of each cell. It is a linear polymer of four building blocks called nucleotides. Nucleotides are distinguished by their base structures known as adenine (A), guanine (G), thymine (T), and cytosine (C).

Enantiomers Compounds that deflect plane polarized light either to the right (clockwise) or to the left (anti-clockwise).

Endoplasmic reticulum Highly convoluted membranous sac continuous with the outer membrane of the nucleus. It is made up of interconnected tubules.

Enzyme A special class of functional protein that catalyses the biochemical reactions during metabolism.

Enzyme specificity Extraordinary ability of the enzyme to recognize a specific substrate to catalyse a specific reaction.

Epimers Monosaccharides which differ in the position of −OH group around a single carbon.

Essential fatty acids Fatty acids that are not present in the body but essential for normal cell functioning and should be taken through diet.

Eukaryotes Organisms with large cells and internal membrane-bound structures called organelles.

Flagellum Hairlike structure attached to bacterial cell, used for locomotion.

Fluid mosaic membrane A model of the cell membrane, which describes the structural and dynamical organization of biological membranes. It is composed of phospholipids that form large planar bilayers. In between phospholipids exist membrane proteins, and the alternating composition of phospholipids and proteins found in membranes has been compared to a mosaic structure. In addition, both components are not fixed in space but can freely move within the plane of the membrane. This 'fluidity' is similar to the fluid or liquid state of water.

Gene A section of DNA comprising a sequence of four bases: adenine, guanine, cytosine and thymine.

Gluconeogenesis Synthesis of glucose from non-carbohydrate precursors.

Glycogenesis Polymerization of glucose molecules to form long-chain polymer of glycogen.

Glycogenolysis Breakdown or depolymerization of glycogen into glucose units.

Glycolysis Stepwise enzyme-catalysed oxidation of glucose with the net formation of ATP.

Glycosaminoglycan Linear polymer of repeating disaccharide units in which one is always either *N*-acetyl glucosamine or *N*-acetyl galactosamine (amino sugar) and the other unit is usually glucuronic acid.

Gyrase An enzyme that relaxes the positive supercoils

in the helical structure of DNA during replication.

Hydrogen bond Bond formed between one of the lone pair of electrons on an oxygen atom and the hydrogen attached to a nitrogen atom.

Inulin Polymer of fructose units.

Iodine number Number of grams of iodine absorbed by 100 g of fat.

Ketose Sugars with ketone group as the functional group.

Ligase An enzyme that seals the gaps formed in lagging strands (between Okazaki fragments) during replication.

Lysosomes Small, spherical or irregular membrane-bound saclike organelles which store an array of hydrolytic enzymes collectively called acid hydrolase.

Macromolecules Large molecules in biological systems namely proteins, nucleic acids, and polysaccharides.

Magnetosomes Specialized inclusion bodies found in the cytoplasm of magnetotactic bacteria and that help orienting the cells in the earth's magnetic field.

Meiosis A type of cell division that occurs in germ cells.

Metabolism The totality of all chemical processes in cells and all living organisms. Metabolism is the chemistry of energy extraction from nutrients and the biosynthesis of the building blocks of life (amino acids, sugars, lipids).

Mitosis Type of cell division that occurs in somatic cells.

Nucleus Membrane-bound organelles which contain the DNA in the form of chromosomes. It is the site of DNA replication, and the site of RNA synthesis.

Organelle Subcellular structure in eukaryotic cells (e.g. plants and animals) providing specialized function within cells. Organelles are separated from each other and the cytoplasm of the cell by membranes.

Pentose phosphate pathway Pathway for the interconversion of pentose and hexose.

Peptide bond Covalent bond created by a dehydration reaction that occurs when two amino acids are positioned so that the carboxyl group of one is adjacent to the amino group of the other.

Peptidoglycans Linear polymer of *N*-acetylmuramic acid (NAM) and *N*-acetyl glucosamine (NAG) linked by ($\beta1 \rightarrow 4$) linkage.

Phospholipid Main lipid component of cell membranes. Phospholipids are a heterogeneous type of molecules composed of glycerol, phosphate, two fatty acid residues, and "headgroups" with different chemical properties.

Phosphodiester bond Bond formed between the 5'-phosphate group of one nucleotide and 3'-hydroxyl group of the adjacent nucleotide within a nucleic acid.

Pili A slender hairlike wavy structure on the bacterial surface which is responsible for conjugation (reproduction) and attachment of virus to the cell surface.

Plasmid Circular loop of DNA in prokaryotes.

Post-transcriptional modification Several structural changes that nascent RNA undergoes to become functional. It is also called post-transcriptional processing.

Protein Proteins are macromolecules made from twenty different types of amino acids. Proteins constitute the active component of cells. Proteins function as enzyme in metabolism, transporters and receptors in cell membranes, hormones, antibodies, and help read, translate and replicate the genetic information.

Pyruvate dehydrogenase A mitochondrial enzyme which links glycolysis and TCA cycle.

Quaternary structure The highest level of organization within a protein complex that describes the number of subunits (individual polypeptide chains) and their interactions.

Reichert-Meissl number The ml of 0.1N NaOH required to neutralize soluble fatty acid in 5 g fat.

Replication fork Point where the two template strands of a double helix unwind to initiate replication.

Ribosome The cellular particles made of protein and RNA subunits that catalyse the synthesis of proteins along a messenger RNA (mRNA) template.

Ribozymes Catalytic RNA molecules.

Salt bridge Bond formed between positively charged amino acids (arginine/lysine) and negatively charged amino acid.It is also called weak electrostatic bond or weak ionic bond acids(aspartate/glutamate) in proteins.

Saponification number Number of mg of KOH required to saponify 1 g of fat.

Simple lipids Esters of fatty acids with the short-chain alcohol (glycerol).

Splicing A process during protein synthesis where the mRNA cuts out the intron sequences and strings together the exon (coding) sequences derived from a DNA template during transcription.

Translation Process by which the information encoded in mRNA is converted into proteins by enzymes, ribosomes and tRNA.

Transport proteins Proteins which transport biological materials, e.g. haemoglobin transports oxygen, lipoproteins transport lipids, transferrin transports iron.

Uracil A pyrimidine base and one of four nitrogenous bases found in ribonucleic acid (RNA). It is part of UTP, the triphosphorylated nucleotide.

Vacuole Membrane-bound fluid-filled space within a cell. In most plant cells, there is a single large vacuole filling most of the cell's volume. Some bacterial cells contain gas vacuoles.

Virus Smallest of all organisms and often not considered alive because they strictly depend on a cellular host organism (bacteria, plant, animal) to reproduce.

Zwitterions Electrically neutral amino acid molecules with no net charge.

REFERENCES

Alan Fersht. (1999). *Structure and Mechanism in Protein Science. A Guide to Enzyme Catalysis and Protein Folding*. W.H. Freeman and Company, New York.

Bhagavan, N.V. (2002). *Medical Biochemistry*, 4th edn. Harcourt/ Academic Press, Masachusetts.

De Robertis, E.D.P. and De Robertis, D.M.F., Jr. (2003). *Cell and Molecular Biology*, 8th edn. Lippincott Williams and Wilkins, Philadelphia.

Donald Voet and Judith, G. Voet. (2004). *Biochemistry*, 3rd edn. John Wiley & Sons, New York.

Erice, E. Conn, Parul, K. Stumpf, George Bruening and Roy, H. Doi. (2002). *Outlines of Biochemistry*, 5th edn. John Wiley & Sons, Singapore.

Hames, B.D. and Hooper, N.M. (2003). *Instant Notes in Biochemistry*, 2nd edn. Viva Books Private Limited, New Delhi.

Harvey Lodish, Arnold Berk, Lawrence, S. Zipursky, Paul Matsudaira, David Baltimore and James Darnell. (2000). *Molecular Cell Biology*, 4th edn. W.H. Freeman and Company, USA.

Klaus, B. Buchholz and Poul, B. Poulsen. (1994). *Applied Biocatalysis*. Boross, L. and Johannes Tramper (eds.). Routledge, USA.

Marsh, E.N.G. (1999). "Coenzyme B12 (cobalamin)-dependent enzymes." In: *Essays in Biochemistry*. 34: 155–172.

Nicholas, C. Price and Lewis Stevens. (1999). *Fundamentals of Enzymology. The Cell and Molecular Biology of Catalytic Proteins*, 3rd edn. Oxford University Press, UK and Europe.

E

Ectoplasm 11
Edman reagent 109
Edman reaction 109, 123
Egg 230, 240, 243–245
Electronegative 220
Emil Fischer reaction 109
Enantiomers 160, 184, 249
Endoplasm 12
Endoplasmic reticulum 13
Entropy 145, 222
Enzyme 1, 2, 4, 15, 17, 18,
    20, 25, 27, 35, 46,
    70–72, 75, 76, 79, 83,
    88, 89, 91, 99, 100,
    119, 129, 132, 137–139,
    140, 142–144,
    147–149, 202, 203, 205,
    210, 213, 215, 216,
    240, 244, 249
Epimers 160, 184
Eukaryotes 7, 8, 36, 73,
    79, 83
Eukaryotic 10, 42
Extremozymes 139

F

F-actin 22
FAD 212, 213
FADH 148
FADH$_2$ 209, 215
Fimbriae 49
Fish 235, 244, 246, 247
Flagella 49, 50
Fluid mosaic model 20, 21

Fructose 3, 4, 154–156,
    163, 164
Furan 162, 163, 185

G

G-actin 22
Geometrical isomerism
    207, 190
Gluconeogenesis 177–180,
    184
Glucose 3, 4, 154–156,
    160–164, 167, 169, 175,
    177, 179–182, 184, 197,
    198, 203
Glyceraldehyde 154, 159,
    184, 185
Glycogen 4, 164, 166, 181,
    182, 185
Glycogenesis 175, 180,
    181, 184
Glycogenolysis 175, 182,
    184
Glycolysis 175, 177, 184,
    203, 210
Glycophorin 173, 185
Glycoside 163

H

Haemoglobin 115, 119, 2
Heterolipids 195
Hydrolase 15

I

Immunoglobulin 119, 120
Inositol 195

K

Katal 138, 150
Ketose 155, 168, 184

L

Lamins 12
Lecithin 195
Linoleic acid 187–189
Lysosome 15, 61

M

Milk 230, 235, 240,
    242, 245, 247
Mitochondria 16, 36, 61
Mitosis 27, 39, 43, 68
Mycelium 37, 41
Myoglobin 114
Myosin 22, 29

N

NAD 211
NAD+ 210, 242
NADH 148, 209, 213, 215
NADP+ 242
NADPH 203
Neuraminidase 58
Ninhydrin 123, 134, 135
Nucleus 12, 29, 36, 43,
    60, 61

O

Oil 187, 194, 235, 236, 238
Okazaki fagments 72,
    99, 100
Oleic acid 187–189

P

Paraplasm 11
Pellagra 229, 241, 242
Pentose 63, 183–185
Peptide bond 2, 35, 88, 89,
    107, 113, 116
Peptidoglycan 170
Pili 49, 50, 60
Plasmalogen 195
Polymerase 72, 73, 76
Primary 2, 65, 67, 68, 75,
    111, 113, 120–122
Purine 64, 67, 85, 93, 95,
    96, 98, 100, 246
Pyran 162, 185
Pyran ring 162
Pyrimidine 64, 67, 85,
    96, 98, 100, 239, 246
Pyruvate 177, 179, 203,
    209–211

Q

Quaternary structure 111,
115

R

Replication 13, 27, 54, 70,
    71, 73–75, 99
Ribose 3, 4, 98, 169, 225
Ribose 5-P 184
Ribosomes 16, 17, 47,
    84, 89
Ribozyme 88, 137, 150, 151
Rickets 236

S

Salvage pathway 94, 96, 97
Sanger's reaction 135
Saponification 192, 207
Scurvy 229, 250
Securin 33
Sex hormones 5
Shine–Dalgarno sequences 85
Sphingosine 197
Starch 4, 156, 164, 185
Stereoisomerism 159
Steroids 5
Supersecondary structure 113, 114
Synovial fluid 219

T

Terpenes 201
Tertiary structure 65, 67, 68, 111, 114, 115, 124
Tetrahedron 220
Thallus 36, 37, 41
Transaminase 130

Transcription 13, 75, 76, 77, 100
Transition state 142, 144
Translation 13, 83, 84, 87, 88, 89, 100, 101
Tubulin 23

U

Urea 139
Urea cycle 130, 131

V

Vacuoles 42, 48, 61
van der Waals 68, 114, 115
Vector 125, 127
Velocity of a reaction 137, 140, 141, 145, 146

W

Waxes 187, 194, 195, 206

Z

Zwitterions 105, 134
Zymogens 91

Made in the USA
Monee, IL
07 July 2026

56550063R00163